THE STAR DRIVE

THE
STAR DRIVE

THE TRUE STORY
OF A GENIUS, AN ENGINE
AND OUR FUTURE

PHILLIP HILLS

BIRLINN

First published in 2021 by
Birlinn Limited
West Newington House
10 Newington Road Edinburgh
EH9 1QS

www.birlinn.co.uk

ISBN: 978 1 78027 668 7

British Library Cataloguing-in-Publication Data
A catalogue record for this book is available from the British Library

Typeset by Initial Typesetting Services, Edinburgh

Printed and bound by Clays Ltd, Elcograf S.p.A

For Maggie

Contents

List of Illustrations		ix
Prologue		xi
Introduction: The Portobello Road		1
1.	Robert Stirling	7
2.	Stirling's Engines	34
3.	Sir George Cayley	66
4.	John Ericsson's Caloric Engine	83
5.	Near Death	104
6.	Street-fighting Man	128
7.	The Giant Killers	138
8.	William Beale and the Free-Piston Stirling Engine	157
9.	The Star Drive	181
Epilogue		212
Select Bibliography		227
Acknowledgements		235
Index		237

List of Illustrations

PLATES

1. Dr Robert Stirling
2. A mysterious machine bought from a stall on the Portobello Road one Friday morning
3. Innerpeffray Library
4. The Edinburgh Engine
5. The Glasgow Engine
6. Sir George Cayley
7. John Ericsson
8. Cayley's glider
9. The Caloric ship *Ericsson*
10. The battle between the *Monitor* and the *Merrimack*
11. The Ericsson Sun Motor
12. A Rider–DeLamater engine
13. A Jost fan
14. Malone's large engine
15. Cross-section of the Philips Stirling engine
16. William T. Beale

17. The free-piston engine as supplied to NASA
18. Schematic diagram of the free-piston engine
19. KRUSTY in the NASA workshop
20. KRUSTY schematic
21. A conceptual drawing of a kilopower reactor on the surface of Mars
22. A NASA imagining of a Mars base supplied with power by scattered KRUSTIES

DIAGRAMS

Diagram of the 1816 engine	42
Conversion of a steam engine (1827)	57
Engraving of the 1840 version of the engine	59
Cayley's first engine	68
The engine in Cayley's brewhouse	80
Schematic diagram of the Ericsson engine (1833)	92
Engine proposed by Rankine and Napier	108
A Rider patent hot air engine	117

Prologue

In February 1990, when Voyager 1 finally left the helio-sphere and was nearly four billion miles from home, it sent back photos of the Earth as seen from the edge of the Solar System: it appeared as a pale blue dot in an immensity of darkness. Never before had humankind been given such a perspective on the planet which we call home. In the far future we may look back on this moment with something of the emotion that Europeans felt in the century after they discovered the Americas. For we can be fairly confident that, if we survive, this will be seen as the beginning of our species' liberation from our planetary origins.

The exploration of the Solar System has been made possible by the greatest concentration of technological innovation our planet has seen, and the leader of that effort for the last thirty years or so has been NASA, the National Aeronautics and Space Administration of the United States. Most people who have any interest in space exploration are aware of this; what few people know is that at the core of the project, and essential to its future success, is an engine which will power all the functions of the spacecraft. It does this by converting the heat of a nuclear fission reactor directly into electricity. The engine was invented more than two centuries ago by a Scottish minister of the Kirk called Robert Stirling – and the Stirling engines which spacecraft will now carry are not so different from the drawings which young Robert made when he applied to patent it in 1816.

In May of 2018, NASA called a press conference to announce the successful test-run of their KRUSTY machine. They explained that KRUSTY was an acronym for an electricity generator to be used for manned bases on the Moon and the nearer planets, and for distant space missions – all of which require a compact, reliable and *small* source of electrical power. The output of solar photovoltaic cells falls off beyond Jupiter and radioisotope generators don't produce enough power to run a dishwasher. (All starships will need a dishwasher.) KRUSTY will provide support for life on satellites and planets and, eventually, propulsion beyond the Solar System.

KRUSTY stands for Kilopower Reactor Using Stirling Technology. The first part of the acronym is largely self-explanatory: a kilopower reactor is a very small nuclear reactor which produces heat sufficient to generate a few kilowatts of electricity. This doesn't sound much, given the scale of the ambition, and a million times less than you would expect from a normal nuclear power station, but it's a lot more than astronauts have been used to until now. But what about using Stirling technology? Technology we know about, but what has it got to do with the name of a town in Central Scotland? The answer, of course, is that the thing uses the engine invented by Robert Stirling, and named after him, to convert the heat of the reactor into electricity.

So an engine on which humanity's greatest enterprise depends was invented by a Scotsman more than two hundred years ago: how astonishing. You could be forgiven for wondering why you have never heard of him or his engine. A few people have, of course, and the engine is well-known to enthusiasts around the world. But if you stop people on the street in Scotland and ask them, you will find they know nothing about it. (I can tell you this is the case: I have tried it.) The book which follows is about that engine and about the people – many of them very remarkable – who have brought it to the point at which it can be installed on a spaceship and expected to run non-stop, without repairs or maintenance of any sort, for tens, perhaps hundreds, of years.

Since the Stirling engine is the only known way of producing an appreciable quantity of long-term electrical power on a spacecraft, and all feasible ways of propelling a starship require electricity, it is reasonable to believe that when a spacecraft from this planet finally makes it to one of the nearer stars, its systems and its propulsion will be thanks to the power from the Stirling engine. Truly, a Star Drive.

The Portobello Road

One fine Friday morning, many years ago, Dick and I walked over to the Portobello Road. It is an undistinguished thoroughfare in north London – undistinguished save on Friday mornings, when there is a street market and it is transformed into something magical. In the market are sold fruit and fish, falafels and hot dogs, old clothes and miscellaneous junk. There is a distinction in the matter of junk, for at the south end it aspires to the status of antiques, while at the north there is no such pretension. Needless to say, the north end is the more interesting.

Having walked the length of the road that Friday, Dick and I stopped by a stall whose merchandise, unusually for the south end, appeared to be of a technical and mechanical nature. On the ground – for the stall was heavily laden – lay a curious little machine.

'What's that?' I asked the stallholder.

'Some kinda model steam engine,' he replied.

'May I lift it onto the stall?' I asked.

'Sure,' he said.

I picked up the device and sat it on a space which the chap had obligingly cleared. It was surprisingly heavy, about 2 feet long and evidently made of cast iron. There were two parallel cylinders joined by a pipe and one of the cylinders had a frill of what appeared to be cooling fins. Rods emerging from the cylinders connected to a shaft, which carried a small flywheel. I gently turned the flywheel and one of the rods went in and the other out.

'I don't think it's a steam engine,' I said.

Dick was looking closely at it. 'I don't recognise it at all,' he ventured. 'What do you think it is?'

I hesitated. It's never a good idea to be too forthcoming if a negotiation is in view and the guy you are to negotiate with is listening closely. Another customer came by and the stallholder transferred his attention.

'I think maybe it's a Stirling engine,' I said.

'What's that?' Dick asked.

I frowned. 'I'm not sure. I've never seen one in the flesh. Only read about them. They were common a century or more ago. I think if that bit' – I pointed to the end of one of the cylinders – 'is heated, the thing will run.'

'On just heat?' Dick asked. 'How's that?'

'I don't know,' I rather lamely replied. This was a shaming moment, for both of us had some experience of machinery. The stallholder turned back to us.

'Well?' he asked.

I took the plunge. 'How much do you want for it?' I asked.

'Two 'undred and forty,' he said, very positively.

I forget what exactly I replied, but I must have expressed incredulity.

The dealer continued. 'This is not just amateur rubbish,' he said. 'You can see it's museum-quality stuff.'

I could indeed. The iron castings, though heavy, were

of a delicacy which it would be difficult now to replicate, and the fettling and machining were faultless.

The dealer continued: 'And I paid a bloke two 'undred and twenty for it yesterday, and I'll be happy to take twenty on it.'

This carried conviction. I had been around antique dealers for many years, and I knew that the serious and successful traders were those who kept their capital working by accepting small margins. In a scurry of mental gymnastics I tried to balance desire with the prospect of domestic reproach at such extravagance and produced a limping excuse: 'I don't have that sort of money on me, I'm afraid,' I said.

The dealer was quick as a ferret. 'I got an agreement with the lady in the shop behind,' he said. 'She can take a credit card.'

There are times when the only decent response to the presence of fate is kindly to acknowledge it, and give in. I gave in and produced a credit card. I signed the chit and the dealer, pleased, put my purchase in a big bag. The bag was very heavy: Dick and I carried it between us until a taxi appeared. When we got back to Camden, my sister asked, 'Is there a funeral?'

I said no, I had just spent money I shouldn't have, and I was going to be in trouble.

We had a good look at the machine and Dick fetched an oilcan. I lubricated all the bits that looked as though they would move. We lifted the machine onto the gas stove in the kitchen and Dick lit the gas under the longer of the two cylinders – the one with a frill on its neck. For a while nothing happened, save that the thing got hotter. There was a burning smell. The cylinder under the gas flame began to turn red and we both became apprehensive. The machine gave a twitch. Then, slowly at first, the flywheel began to turn. The piston rods moved in and out and the speed gradually increased. It was soon spinning merrily and starting to shake – though not much, for the flywheel was evidently well-balanced. It ran faster and faster and eventually we felt we really ought to turn the gas off.

Dick leaned back and wiped his hands with a rag. 'Well, so that's a Stirling engine. How does it work? Why does it go round?'

'I've no idea,' I said.

I did meet some domestic resistance, but not much. At least the engine was small, and not smelly, like some of the things I brought into the house from time to time. Having bought a Stirling engine, and shown it working to some of my friends, I was moved to find out something about it, especially as regards that question of how it worked, which was what everyone wanted to know. I bought a few books and borrowed others. The internet was in its infancy at that time, so instant answers were not available as they are now. Over time I came to some sort of understanding of why the wheel went round and, being a lazy sort of person, was more or less satisfied. Some of the books about it had been written by amateurs and were reasonably accessible to a person of my level of understanding, but the more technical works were well beyond me. It seemed that most folk who knew a lot about Stirling engines would, at the drop of a hat, go off into discussions of thermodynamics where I couldn't follow them – mainly because their disquisitions were couched in mathematical equations and assumed a knowledge of physics which was well beyond my first-year university course of thirty years before.

From time to time, over the next decade or two, I would pick up one of the books again, or buy another one. The maths didn't become any more transparent, but I did get an inkling that there was more going on than just mechanical curiosity. There was some mystery about the engine: why had it been so popular and why had it vanished? There were thousands of them in the nineteenth century but they disappeared in the twentieth. And there seemed to be something which nobody could quite explain about how it worked: what was it? And what had it to do with the town of Stirling, if anything? I had read that the engine was invented by a man of that name, in Kilmarnock, a town in Ayrshire of which I knew next to nothing, save that Rabbie Burns, Scotland's national poet and bad role-model for Scottish boys, had lived nearby and spent a lot of time there drinking with his pals.

Now, I am Scottish, born and bred, and members of my family have lived in Scotland for five hundred years that I know about. (There were some very bad ones, which is why I know about them.) Yet until fairly recently I had never heard of a Stirling engine. This in a small, boastful country which is much given to reminding the world of how many famous people have been born here. Nobody I knew had heard of a Stirling engine either – and I know a lot of people.

Since the mathematics seemed to be a barrier to my fully understanding how the thing worked, or to not-quite-understanding how it worked, I decided to find out what

I could about the people who had been connected with it. They turned out to be astonishing – and had some even more astonishing connections. Stirling himself was a bit of an enigma: he had left no letters and lived a perfectly blameless life as a minister of the Kirk of Scotland, so little was known about him (except for the Great Strathbogie Controversy, of which more later). The only research which had been done on his life had been done by people who knew next to nothing about Scots history or culture; especially Scots culture in that crucial time at the start of the nineteenth century when Stirling invented his engine. So there was work to be done.

Once I had started down that road, the Stirling associations came fast and thick. First there was Sir George Cayley, who had invented an aeroplane before there were even railways. Then came John Ericsson, a Swedish American who became a national hero of his adopted homeland (or at least of the Federal part of it in the Civil War) and whose fellow-Swedes sank the greatest warship ever built (only in make-believe, but they could have done it for real). One chap, John Malone, who is an important part of the story, was notorious as a knife fighter before he became an engineer, while another, H. Rinia, allowed the Nazis to steal the engine, knowing they would make nothing of it, and then reinvented the thing. But it wasn't until I read about NASA's New Horizons probe passing Pluto that I came across the name William Beale, an inspired and brilliant American who for tough-minded saintliness is up there with Stirling, and whose take on the engine was to send it to the stars. The little engine I had bought on the Portobello Road was throwing up a lot of questions.

CHAPTER 1

Robert Stirling

The summer was bad that year, and it was bad all over. There were frosts in the spring, and in some places there was snow in June. There was frost again in August and the crops rotted in the fields. Farmers who had poor harvests thought themselves fortunate, for most had no harvest at all. The wheat crop failed; so did the potato. Even the oats, that hardy cereal which had evolved in northern climes, were mostly not worth the trouble of raising from the mud in which they lay. The sun was rarely seen, and when it was, it was blood-red. The rain never seemed to stop. It was the worst summer, by far, in nearly two thousand years.

The date was 1816 and the whole of the northern hemisphere was affected. The weather had been bad for some years past: most people had eaten their reserves of corn and there was nothing left. Famine crawled the land: across Europe, Russia, China and North America people and livestock died. More in some places than in others, but everywhere there was distress. The cause, as we now know, was a series of volcanic eruptions from 1812 onward, culminating in the eruption – or, rather, the disintegration – of Mount Tambora in what is now Indonesia. It was the biggest blast in recorded history

– bigger by far than the sum of all the nuclear explosions since 1945. An aerosol of sulphur compounds was hurled into the stratosphere, the sunlight was dimmed and global temperatures dropped by about one degree Centigrade. That is a lot more serious than it sounds.

In an age of widespread belief in supernatural agency, people understandably applied to their god or gods for relief, though none seemed disposed to help. In the Laigh Kirk in the town of Kilmarnock in Scotland, the congregation looked on the bright side and thanked the Lord that their church had been rebuilt a few years before and at least the roof kept out the eternal rain. But, being good Calvinists, they regarded the foul weather as divine retribution for their – or at least for their fellows' – sins. Such views were becoming less common among the more educated part of the populace, due to the influence of Enlightenment rationality in the preceding half-century. But even in the less-evangelical churches, a great many people clung to the notion that by amending their evil ways or, more commonly, by persuading their fellows to amend *theirs*, they might induce the deity which governed the universe to suspend the operation of causality in their favour.

In this they showed a certain lack of perspective, which is not uncommon in religious congregations of all denominations. The idea – that the deity might be persuaded to undo the effects of a few million tons of Mount Tambora dispersed in the stratosphere, in return for improved behaviour on the part of the townsfolk of Kilmarnock – seems disproportionate from our standpoint. It was not, apparently, from theirs. Prayers were offered up for better weather and promises of amendment were made.

In fairness to the Almighty, it must be admitted that the Kilmarnock folk didn't keep their side of the bargain, so they couldn't really complain. No doubt the promises were sincere at the time they were made, but the Minute Book of the Laigh Kirk session for the following year suggests otherwise. It shows the number of children born out of wedlock to have been slightly greater than before. (The Minute Book implies that there was little sin in Kilmarnock other than fornication. Since this would be viewed more tolerantly today, perhaps the Almighty's reluctance to reverse the Mount Tambora event is understandable.)

In 1816 the Kirk of Scotland was not short of ministers ready to attribute failed harvests to human sinfulness, but their number was diminishing with the entry to the profession of young clergymen who had imbibed the philosophical and religious precepts of the later eighteenth century. The young man who had recently been called to the second charge in the Laigh Kirk was one of the latter; his name was Robert Stirling. His tenancy of the position was proposed in June and ratified in September. By the time he took up his duties, the dire condition of the country had become very obvious. We know from his actions and his later sermons that his inclination was rather to address a problem than to blame it on divine retribution for supposed immorality. Even so, what he did next was rather surprising: a week after his ordination he lodged an application to patent a device which would later be called a 'heat exchanger'.

The purpose of the heat exchanger was to economise in the use of coal in furnaces by using the waste heat from the furnace flue to heat the air coming into the fire. (A fire which is supplied with hot air burns hotter than one which has to heat the incoming air.) It did this by doing what heat exchangers have done ever since: the cold incoming air is made to flow past the outgoing gas, smoke, etc. with only a thin wall between them. The heat is conducted through the wall from the latter to the former. It works both as a heater and as a cooler, depending on your point of view. The radiator of your car is a heat exchanger in which the heat from the engine cooling water passes through thin tubes and is transferred to the air which blows through it. Your central heating boiler passes the heat from burning oil or gas to the water which heats the radiators. The latter are also heat exchangers, since they transfer the heat from the water to the air of the room.

The Stirling heat exchanger invention was later to become important in the development of the steel industry and on that account alone the new minister of Kilmarnock kirk may lay claim to a place in industrial history. But it is with the second part of the patent application that we are concerned, for it describes an engine which, using modifications of his invention, might be applied to moving machinery. The application includes a drawing of the engine and from this, as well as the text of the application, it is apparent that Mr Stirling had more things in mind in the years preceding his ordination than theology and divinity.

Robert Stirling knew about bad harvests, for he had been a farm boy. Cloag Farm, where he was born, in the parish of Methven, near Perth, lies on the south-facing slope of the Grampian Mountains, overlooking the valley of the

River Earn. The farm is still there today, as is the grey stone farmhouse which is almost certainly his birthplace. It is a fine if modest stone house set in tall, mature timber; prosperous but without ostentatious signs of affluence. The Stirlings were an extended family of middle-class agriculturists whose farms were strung across that southerly slope from Stirling to Perth. They were mostly tenant farmers, but nonetheless possessed of capital sufficient to stock and work a farm – which put them well above the labouring classes, both socially and economically. And the family was cohesive, each member being aware of the number and extent of his or her relations. Among Lowland families there was no formal clan structure such as that which prevailed in the Highlands (which began only a few miles north of Strathearn), but the Scots were notorious for their pride in, and adhesion to, family.

The land which his father farmed had been mostly forest as little as a century earlier, and what had not been woodland was bog. But from the early eighteenth century, a spirit of inventiveness and a habit of rational exploitation of possibilities of improvement had led to a huge increase in the fertility and productiveness of the land, and a rising population in growing towns and cities provided markets for agricultural produce. It was a virtuous circle which made possible the developments in science and technology which were shortly to produce the second stage of the Industrial Revolution.

By the time Stirling was born in 1790, Cloag Farm had introduced new crops, some from overseas, which had been selectively bred to produce higher yields than the old varieties; methods of tillage had changed and machinery had reduced the need for human or animal muscle in planting, harvesting and processing. The machinery was mainly made of wood and would be barely recognisable as machinery today, but the principles on which it operated were coming to be understood throughout the farming community. Stirling's grandfather was reputed to have invented a threshing machine, probably horse-powered, which separated the seed from the stalk and replaced the ancient flail. Machinery was driven by water, wind or animal power. (The steam engines which had become common in industry by the time of Robert's birth were too big and too expensive for farm or village use.) In 1790 in the village of Methven, which lay at the foot of the slope below Cloag, there were several mills. The term 'mill' may be misleading here, implying as it does buildings of the dark satanic variety. Village mills were not like that: they tended to consist of cottages, in one end of which the operative pursued his or her craft, while the family lived in the other end – an arrangement sanctioned by tradition, but with cattle replaced by loom or spinning wheels. But they did contain machinery, of however simple a sort. The two-handed spinning wheel introduced in 1770 doubled the productivity of the spinner, thanks to a device which, though easy to make and to handle, introduced people to the notion that mechanical innovation might greatly increase their prosperity. As a consequence, young Robert grew up in a society whose fabric was shot through with the idea that life could be improved by the application of energy and ingenuity.

At the age of fifteen, young Stirling went from Cloag to Edinburgh – probably on foot, a matter of 50 miles or so.

(In those days, most people did not stray far from home for most of their lives; and, unless they were wealthy, those who did, walked.) That he went equipped with a high degree of literacy, we know. He probably attended Perth Academy, which is some 5 miles distant from Cloag: a fine school open to boys of any social class, with a progressive curriculum which, besides the traditional ancient languages, taught contemporary subjects such as scientific navigation. Alas, the archives of Perth Academy do not go back far enough, so we know little of his schooldays. But we do have some evidence of Robert's reading from a most curious institution, which today is little changed from when Robert visited it to borrow books.

About 15 miles from Methven lies Innerpeffray Library. This is no ordinary urban library. It lies deep in Perthshire farmland, not far from the town of Crieff. At the end of an avenue of trees stand some ancient buildings which house the oldest lending library in Scotland. There have been church buildings on the site since at least the fourteenth century, but none used as such after the Reformation. From mediaeval times the land had been in the hands of the Drummonds, another of the extensive Lowland families, but much richer and grander than the Stirlings, counting peers, an earl and later a duke among their number. David Drummond, the third Lord Maddertie, who died in 1692, had left instructions in his will that his library remain at Innerpeffray, and two years later a charity, the Innerpeffray Mortification, which is still in existence today, was established to care for the collection. Since the books were almost all to do with astrology, religion, witchcraft and demonology, this was scarcely a great contribution to human understanding, but by the standards of the time they were regarded as scholarly. Things improved with the addition of the collection of the Rev. Robert Hay Drummond half a century later, and by the time Robert Stirling was of age most of the works of scholarship and science of the later eighteenth century were available. And, astonishingly, given the period, they were available to be borrowed by local people of little or no social standing. All that was required was that the borrower sign the Borrowers' Register, giving their name and place of residence. That Borrowers' Register is in existence today and may be inspected on application to the librarian.

In the volume for 1806, when he was sixteen years old, the Register carries Robert's signature and his place of residence, Cloag, to evidence his borrowing first *Buffon's Natural History* (presumably not all thirty-six volumes; more likely John Leslie's translation, for which see later), then Gibbon's *Decline and Fall* and, later, Robertson's *History of America*: scarcely convenient additions to the college curriculum. These works are no light literature: Robert Stirling was plainly a serious-minded young man. So, though there is little record of Robert's early life, we know that he went up to Edinburgh University equipped with an impressive spread of information.

The Edinburgh to which Stirling travelled was an extra-ordinary place. At the beginning of the century it had been a narrow, strait-laced parochial town, introverted and governed by an exclusive and self-serving town council. By the century's end it would become a society which people, not its own, would compare seriously with the Athens of Pericles or the Florence of the Medicis. The university, whose professors were paid a pittance and depended on students' class fees for most of their income, would come to have an international reputation in philosophy, literature, history, medicine and the sciences. The great Dr Johnson, who did not love the Scots, said, 'A man may stand at the cross of Edinburgh and shake the hand of twenty men of genius.'

In compiling a biography, the writer commonly seeks out the writings of the subject or, failing those, written evidence from his or her contemporaries. As regards scientific or technical advance, it is generally possible to trace the antecedents of the innovation through publications in scholarly journals. In Robert Stirling's case, we have almost none of the former and little enough of the latter: with regard to both, we must rely on circumstantial evidence and proceed where possible by way of inference – for Stirling left only his sermons and his patent applications – and his great invention was perfectly unique, owing little to antecedents. That is not to say that we cannot identify some of the influences which contributed to Stirling and his works. Of those influences, easily the greatest is the milieu in which he found himself on going to Edinburgh and, later, to Glasgow. Each in its way was different from the other and each was hugely important. We will deal with the scientific background first; the ecclesiastical we shall come to later.

Edinburgh was literally extraordinary. There are few examples of small cities which, over quite a short period, have had a comparable influence on the history of human thought. Two, Athens and Florence, I have mentioned above. It's hard to think of a fourth. Over little more than half a century, the town had gone from hosting the Jacobite Pretender and his Highland Host in a small, insanitary jumble of high-rise tenements clinging to the spine of the castle rock, to a city of planned spaces and elegant neoclassical houses. The town on the rock was still there and still insanitary, but the people who mattered in 1805 would no more have thought of hosting a rebellion against the Crown than of promoting the emancipation of Roman Catholics. Times had changed and so had the people.

That said, the city still thought of itself as a small town and it was possible for a citizen to know and be known to almost everyone of any consequence. The move to the New Town had taken most of the upper classes north to the environs of George Street, but in the lands of the High Street and the Canongate there remained enough of the old aristocracy to leaven the social mix. It was a literate and loquacious society: one in which an ability to use words, written or spoken, was the greatest social asset. It was also a society greatly given to gossip, and gossip was the main means by which news was promulgated.

A student going up to the university for the first time would get much of his information from the gossip of townspeople and his fellow students. All the professors were well known to both. Indeed, such were the eccentricities of many of the professors that they formed a constant source of interest and anecdote. And they were a pretty astonishing bunch. William Cullen and David Hume had died a generation before, but their doings were still living as legend. Hume especially, since as a reputed atheist he was at odds with the most important value systems of the time – and as a genial, friendly, witty, immensely well-informed bon viveur he was more than acceptable to everyone who cared for civilised society. As a philosopher he is today recognised as one of the greatest and the reputation which Edinburgh acquired as the Athens of the North owed something to Hume's status vis-à-vis Plato and Aristotle. Cullen is important to our enquiry, for among his many and varied interests he included a study of heat. In 1756 he produced ice by artificial means; probably the first time this had been done.

Of the academic characters whose death closely preceded Stirling's arrival in Edinburgh, William Robertson and Adam Smith were both outstanding figures, each contributing to the fabric of the legend, but of more direct relevance to Stirling were Joseph Black, who died in 1799, and John Robison, who did the same in 1805. Robison, who was professor of natural philosophy until his death, in early life was far-travelled; among his voyages was one to Jamaica in his capacity as a member of the Board of Longitude to test the accuracy of Mr Harrison's chronometer. He spent time in St Petersburg as a teacher at the naval academy of Catherine the Great. Like Joseph Black he was a friend of James Watt's, encouraging empirical investigations into the nature and behaviour of heat and opposing the theory-dense hypotheses of the continental philosophers. He was appointed to the chair of natural philosophy in Edinburgh in 1773, a post which he held until his demise.

Joseph Black was one of the giants: a polymath who contributed to knowledge wherever he turned his attention. Anatomist, chemist, physicist, physician, he brought chemistry out of the Dark Ages by his adhesion to rigorous experimental method and measurement; his discovery of latent heat (that required to effect a phase change in the nature of a substance) laid the foundation stone of thermodynamics and his discovery of carbon dioxide was one of those which brought chemistry into the modern world.

John Playfair had been professor of mathematics since 1785, and when Robison died in 1805 he moved up to the chair of natural philosophy. (The salary of the latter, at £100 per annum, was double that of the professor of mathematics.) As professor of what we would now call physics, he was in a position to publicise his views on the conservation of energy, as against those of Gottfried Leibniz's *vis viva*. The details of this controversy, which ran throughout the later eighteenth century, need not concern us here, but it is worth noting that it involved the relationship between work and heat.

In 1805, round about the time Robert Stirling was arriving in Edinburgh, Playfair was succeeded in the chair of mathematics by the extraordinary John Leslie. Of humble origins – born in Fife, father a joiner – Leslie appears to have been that rarest of prodigies, a mathematical autodidact. After the elements of education in a village school and a spell as a shepherd, he entered the University of St Andrews at the age of twelve or thirteen. Having finished the Arts course in 1784, he entered Edinburgh University, where he studied divinity. He did not graduate but spent the next two years as a private tutor, some of it in Virginia. In London in 1793 and being poor, he embarked on a translation of Buffon's *Natural History of Birds*. No small undertaking, this was published in nine volumes, was an immediate success and provided cash to finance his studies in, among other things, the nature of heat. In 1804 his *Experimental Enquiry into the Nature and Properties of Heat* was published. It provided the solid basis for his later career – and it also gave rise to a tremendous squabble.

Succession to a chair was usually uncontroversial, but in Leslie's case opposition was raised by members of the Church on the grounds of his apparent atheism. At that time the Church of Scotland was split into two rival factions: moderates and evangelicals. The division was complicated and infested most decisions of the General Assembly of the Kirk. It would eventually lead, nearly forty years later, to the greatest schism of the Church since the Reformation. The moderate faction, being mostly Tories, took exception to Leslie's liberal politics and sought to have his candidacy for the professorship set aside in favour of one of their own undistinguished members. They claimed to have found in a note to the main text of the *Experimental Enquiry* an implication of scepticism. They had also discovered a long-disused article in the council's regulations governing the appointment of professors which stipulated that the Church's opinion must be sought before making an appointment.

Despite being a pious Christian, Leslie had written approvingly of David Hume's theory of causation. (Hume said that all we can know of the causes of natural events is by experience of the invariable succession of cause by effect. This was thought shocking because it left God out of the loop.) Ample evidence was given of Leslie's orthodoxy in matters of belief and of his exemplary personal life. The General Assembly decided by a large margin in Leslie's favour, the motion was lost, the matter was dropped and Leslie was confirmed as professor of mathematics on £50 a year.

The relevance of this to our fifteen-year-old going up to Edinburgh University may not be immediately apparent. Its pertinence is this: the case against Leslie was conducted in public; it was of great interest not only to the members of the Church and to the university, but also to the townspeople, most of whom took a deep interest in all the goings-on of both. A new student entering the university for the first time cannot have been unaware of it and probably heard more about it than he wished, for it was the talk of the very talkative town. A student like Stirling, having both a religious vocation and an interest in the physical sciences, cannot have escaped it. As an incentive to the study of both, it could not have been bettered.

Of Stirling's years in Edinburgh we know little, save that at the end of them he does not appear to have graduated – not an uncommon occurrence since graduation incurred expense and was not a requirement either for subsequent study or for ordination. It can be assumed that Stirling's attendance at classes and his completion of coursework were satisfactory, for he was to be accepted by Glasgow University as a student of divinity for another four years. We do know, from John Playfair, the then professor, what he studied in the natural philosophy course of his Arts degree: it covered the sciences of mechanics, hydro-dynamics, astronomy, optics, electricity and magnetism – and, presumably, heat, given the proximity of John Leslie in the mathematics chair.

It was quite common in the Scotland of those days for a student to study for an Arts degree in the one university and then proceed to the other for what we would call a postgraduate degree. In both theology and the sciences, a great deal of the information the student acquired would derive from the professor's lectures – and, since professors differed greatly in their views, a change would bring a welcome perspective. Stirling spent four years in Glasgow and then returned to Edinburgh for a further year. Again we know little of what he did in either university and are dependent on inference for our data.

The Glasgow to which Stirling went was tiny by modern standards but modern compared to Edinburgh, which by the year 1810 was already ancient. Glasgow had grown rapidly in the preceding century, mainly on the proceeds of foreign trade and the textile industry. Its college, which is no more, was a handsome course of buildings abutting the High Street, surrounded by the bustling of a very lively little city. A short distance away was the River Clyde and the shipping which carried the produce of the city and the surrounding countryside to the colonies, mainly in North America and the Caribbean. (The shipping was still small enough to navigate the river. It would soon outgrow it.) There was a canal, recently completed, which stretched all the way from the city across central Scotland, so that imported goods destined for continental Europe could be carried economically. The roads were miserable even by local standards, though turnpike trusts (which had the right to collect tolls for the maintenance of the highway), where they were active, helped remedy the deficiency.

What we would call transport infrastructure was improving rapidly, though railways were still some decades in the future. Improvement was needed: not only for the movement of goods, but for the transport of materials. Of the latter, the greatest by far was coal. The Central Belt of Scotland was well-supplied by nature with coalfields, and in many places the coal was close to the surface and could be mined using relatively primitive technology. Coal was used for heating, as an export commodity, and for the nascent iron industry in the West. For what would come to be called the Second Industrial Revolution was getting under way. (The First Industrial Revolution had been based on the manufacture and distribution of textiles; the Second was about coal, steam, iron and eventually steel – and produced the ships and railways which would result from their combination.)

It was an immensely interesting time to be alive, especially for a young man who had been studying the nature of heat – for the greater part of the new industrial development was based on what might be done using the energy supplied by burning coal. Factory production of textiles was susceptible to mechanisation and machines required a mover: first the waterwheel and then the steam engine. Coal mines had been kept dry for nearly a century by Newcomen engines and there were many at work in Scotland. Designed by Thomas Newcomen, they would be more accurately described as atmospheric engines, for the steam was used only to push the air out of a cylinder so that, when it condensed to create a vacuum, the pressure of the atmosphere would cause a piston to move. Such engines were the size of a four-storey house, expensive to build, expensive in coal to run, and definitely not portable.

The Newcomen engine had been improved by James Watt's separate condenser, and then improved again and again by subsequent inventions. By 1805, high-pressure steam had made its appearance: in this development, the steam was at such high pressure that it simply pushed the piston directly. Though dangerous, it could be made small and portable and was economical in coal. Compound steam engines would improve it even further and establish a basic design which would last until the internal combustion engine made the steam engine redundant.

It would have been reasonable to expect that Stirling, being interested in engineering, would have done two things: he would have abandoned a career in the Kirk and he would have sought to make his mark by contributing to the dominant machine technology of the era, the high-pressure steam engine. Stirling, curiously, did neither. He continued his studies in divinity until, in 1815, he was licensed by the Presbytery of Dumbarton as a probationary minister– and he invented an engine which presented an alternative to hegemonic steam: an engine which converted the heat from a coal fire into motion by expanding and contracting air.

The Church of Scotland, in which Stirling became a minister, had been in existence for about two hundred and fifty years. Born of the Reformation in the sixteenth century, its basic beliefs differed little from those of the Roman church which it replaced, but in the matter of Church government the two could not have been further apart. The Church of Scotland – or the Kirk, as it was known – was an early attempt at ecclesiastical democracy, based on the equality before their maker of all Christians and the right of all to have direct access both to holy writ and, through prayer, to the attention of the Deity. The units of Christian organisation were the Congregation, the Session, the Presbytery and the Synod, to which all might in principle make representation, with appeal to an annual General Assembly composed of ministers and lay representatives.

This was a welcome change from the sclerotic and hierarchical Roman church, but in the seventeenth century the whole thing went horribly wrong, and the Kirk fell into the hands of scoundrels and fanatics who produced more than half a century of religious warfare, cruelty and tyranny. In fairness, we should say that the Kirk was not alone in this; most of Europe endured a long period of religious wars in which professing Christians behaved like barbarians to their fellow religionists in the name of peace and love, and on the basis of perfectly irrelevant doctrinal differences.

By the early eighteenth century, none of the differences had been resolved, but persecution had gone out of fashion and people became more tolerant of difference. With a view to preventing a recurrence of the killing times, the government of the Kirk was drawn closer to the structure of secular society. Since the latter had persisted since time immemorial, it was thought that this would provide a much-needed stability. The legal and economic basis of Scottish society was the ownership of property, by which was meant land. It was a relic of the feudal system brought by the Norman conquest: most of the Scottish land mass was owned by a small number of proprietors who rented the land to those who cultivated it, usually by way of intermediate tenants such as the Stirlings, who formed a rural middle class. The instrument by which the Kirk was to be tamed was the introduction of patronage, whereby, though the final election of a minister lay with the representatives of the Kirk, the choice would be made from among persons nominated by a patron, who was usually the ultimate owner of the land of the parish. (Needless to say, the change was regarded by many members of the Kirk as a scandalous innovation and some of them left to form the first of the many schismatic congregations which contribute so greatly to the delight of scholars of church history.)

In 1816, the Commissioners for the Duke of Portland proposed the young Mr Stirling as a suitable minister for the second charge of the Laigh Kirk in Kilmarnock. The Duke, William Henry Cavendish-Scott-Bentinck, was a very large landowner in both Scotland and England, a Member of Parliament and for many years highly placed in government. It seems unlikely that the Duke himself would have been aware of the existence of so lowly a creature as Robert Stirling, but the young man must have come to the favourable notice of his functionaries, the Commissioners, who were important persons in the locality of Kilmarnock. (It is not impossible that the Duke knew of Stirling, for he was an active supporter of technological innovation and agricultural improvement. He owned large estates around Kilmarnock.)

Becoming a minister of the Kirk of Scotland was not a matter for the faint of heart: the process was long and daunting. Getting a PhD in a modern university is a walk by comparison. The aspirant had firstly to qualify for a Master of Arts degree at a Scottish university: a four-year course in which the student had to attend classes in Latin, Greek, mathematics, natural philosophy (physics), logic and metaphysics, and law. (I can speak from experience here, having taken logic and metaphysics 1 at Glasgow University, a course which had not changed significantly, at least as regards the logic, since Stirling took it in 1806 – or, for that matter, since William of Ockham five hundred years before.) The Arts course provided a broadly based curriculum designed to provide the student with a high level of general knowledge before proceeding to a specialism. As an institution it has stood the test of time, for some Scottish universities still offer such a course today. Having completed his Arts degree, the aspirant minister then undertook the degree course in divinity. This was no small matter, and we know that Robert attended Glasgow University for four or five years from 1809. Among other things, he would study theology, biblical texts, biblical exegetics, homiletics, hermeneutics, church history, moral philosophy, Hebrew and Greek. He would emerge understanding a variant of the English language which was opaque to ordinary folk – though comprehensible to a good many lay church members. Once ordained, as a minister he would have the difficult job of making his preaching comprehensible to the ignorant while satisfying informed critics of Reformation theology among his congregation. There were surprising numbers of the latter.

Attending university was not the end of the matter for the prospective minister. He then must undergo examination by a Presbytery which, if satisfied – and they were not lightly to be satisfied – would license him as a probationer. Then, if he were presented to a parish by a patron, he would have to undergo another theological obstacle course set by his prospective parishioners before being accepted as a minister. The Kirk of Scotland was no pushover.

The Presbytery of Dumbarton, having subjected young Stirling to an exacting oral examination, licensed him as probationer on 26 March 1816. A little later, the Commissioners for the Duke of Portland presented him as a candidate for the second charge in the Laigh Kirk in Kilmarnock, which had fallen vacant. His ordeal wasn't over: he then had to undergo examination by the Presbytery of Irvine, the body into whose jurisdiction the Kilmarnock Kirk fell. This he survived; then he had to preach a sermon to his flock-to-be in the Laigh Kirk. They approving, he was ordained on 19 September. On the 26th, he lodged his application for patent in respect of his invention of a heat exchanger and engine. The weather had been very wet almost all the year.

———

The Kilmarnock to which our man went in the late summer of 1816 was a town which was remaking itself.

A generation before, it had been a jumble of rickety, muddled, tumbledown buildings packed closely round the muddy streets of its centre. But for some years past, wealth had flowed into the town as its citizens profited from their trade and industry. Old houses were being pulled down to be replaced by grander, more spacious houses, shops, mills, warehouses and public buildings, and by 1800 the town could boast five churches, fifty ale houses, two public schools and many private, a post office, a print works and a public library. Some of the streets were even paved. Mechanisation of the traditional textile crafts had arrived in 1743 with a water-powered woollen mill; silk weaving arrived in 1770 and cotton in 1780. It may give some idea of the scale of this trade to know that by the year 1800 the town had more than a thousand weavers in the muslin manufactory alone.

In the later eighteenth century, the Turnpike Acts had allowed the town to build good roads for the export of its goods, first to the smaller Ayrshire ports and later to Glasgow. In 1785 large estates around the town were purchased by the Duke of Portland and in 1812 he built the Kilmarnock to Troon railway (for horse-drawn vehicles only, mainly carrying coal. The rails were originally made of wood, later of cast iron. Neither material was capable of supporting heavy vehicles.)

The public enthusiasm for mechanical inventions was amply indulged by the goings-on in the Portland estates just outside the town. That same year of 1816 in which our man came to Kilmarnock was witness to the most exciting exhibition of mechanical inventiveness hitherto seen in Scotland, even though, as usual that year, it was raining.

The Duke of Portland, whose Commissioners had presented Stirling to the living of the Laigh Kirk, was a high-minded aristocrat who found his vocation in public service: as a member of government and as the owner of vast estates in which he sought to promote improvements. He was, moreover, keen to utilise technological advances to promote those ends. The Portland estates were sitting on a coalfield, and mines had been sunk which were productive of coal far in excess of what could be consumed locally. Exporting the black gold was, however, difficult, for coal is heavy and the roads, though better than they had been, were incapable of carrying coal trucks. The solution to the problem was to construct the railway from Kilmarnock to Troon, a small port on the adjacent coast. It was much cheaper to build a railway than to make a road: in a nearby iron works, rails were cast as interlocking iron plates in which the wheels of horse-drawn wagons might run. It was a relatively simple device which worked well enough, though there could be a problem if the plates were not properly laid or the load was too heavy, for cast iron is a brittle material which is apt to fracture when stressed.

The high-pressure steam engine had by 1815 been used to propel a railway locomotive successfully. As long ago as 1804 such an engine, built by the extraordinary Richard Trevithick – a giant by any measure – had pulled 10 tons of iron and seventy men a distance of 10 miles on the Merthyr Tydfil tramway. At Killingworth in Northumberland, George Stephenson had built a steam locomotive which had also been successful and a second, improved version. The Duke of Portland had the latter brought to Kilmarnock in 1816 and set it to work on the

Portland railway amid public trepidation and rejoicing. Alas, the experiment was not a success, for the locomotive was so heavy that the railway plates broke and the attempt was abandoned. There was huge publicity, though, of which the minister of the Laigh Kirk and his friend Mr Morton were undoubtedly well aware.

Thomas Morton was in his early thirties when Stirling came to town and he was already one of the more prominent citizens. Another autodidact, his life is a further example of the versatility of the Scottish intellect of that time. With almost no formal education, from the age of ten he worked for his father, who was a brickmaker. Apprenticed at fifteen to a wheelwright, he became an expert turner and adapted a lathe to turn ovals. (An oval was the preferred section for the spindles of a cartwheel, being lighter than a round spindle and stronger where it mattered. It is not easy to turn on a lathe, which generally makes things round.) By age twenty-three, Morton had a workshop in Morton Place in Kilmarnock and was in business on his own account. There he machined bobbins for the local carpet weavers and, having become interested in the machinery which wove the carpets, he devised a loom which by means of pins allowed the weaving of the pattern to be automated. This invention was taken up by all the local carpet weavers (he never patented it), who commissioned him to adapt existing looms and to construct new ones.

His prosperity established, Morton turned, as one would, to astronomy. Very much in the spirit of the time, he first built himself a tower to use as an observatory and then proceeded to manufacture the telescopes he would need to view the stars. This meant making the lenses, as well as all the other parts of the telescopes, from scratch, out of pieces of glass and bits of brass: skilled work drawing on a number of different kinds of expertise. (There is a beautiful Morton-made telescope in the collection of the National Museum of Scotland.) Given their shared interests, it is not surprising that Stirling turned to Morton for workshop space in which to develop his engine. Morton built him a workshop behind his observatory tower which Stirling occupied for years, even after he had left Kilmarnock for nearby Galston parish.

We have no evidence of Stirling's desire to study celestial bodies prior to his going to Kilmarnock but, given that the curriculum in the natural philosophy segment of his Edinburgh Arts degree contained astronomy, it would be no more than natural that he should join his friend Morton in his observations – and in the construction of the telescopes. Apart from anything, working in brass must have been a pleasure after filing and machining iron and the steels then available. There is a story that Stirling's favourite telescope lens was ground from the bottom of a glass bottle. This is not improbable, for the size and shape of glass he probably worked with would have suited the ends of bottles and the bull's-eyes which were a by-product of blown-glass window panes.

Kilmarnock was no mean town. It was a place of opportunity for a lively, intelligent, scientifically minded man of God. But it was not for the faint-hearted: its populace was rumbustious and sometimes violent, and they were apt to express their opinions forcibly. The Laigh Kirk, in particular, attracted a vociferous attendance. It was a large church and two clergymen were considered needful to minister to its extensive congregation. The minister of the first charge, whom Stirling joined in 1816, was a Dr Mackinlay, who had been in post since 1786. The prospect of joining Dr Mackinlay must have been a daunting one, for Mackinlay was one of the evangelical party in the Kirk, and Stirling, being young and educated, was a moderate. (The two parties were delightfully known as the Auld Lichts and the New Lichts respectively.) That the congregation didn't like moderate ministers is shown by the fact that on the appointment of a moderate predecessor to the place which Stirling was to occupy the congregation rioted so violently that three of them had to be whipped through the town of Ayr and imprisoned for a month.

And more: Mackinlay, on his accession, had been the subject of a lengthy and irreverent ballad entitled 'The Ordination' by a local poet. It's a fairly scurrilous work, which takes shots at a lot of local people but gives high praise to Mackinlay and those of his hellfire brethren. One verse will give something of its flavour:

> Curst 'Common-sense', that imp o' h—ll*,
> Cam in wi' Maggie Lauder
> But Oliphant aft made her yell,
> An' Russell sair misca'ed her;
> This day Mackinlay taks the flail,
> An he's the boy will blaud her!
> He'll clap a shangan on her tail,
> An' set the bairns to daud her
> Wi' dirt this day.

*Such words as 'hell' were thought too offensive to be printed.

'Common-sense' was a term used to describe the views of the moderate churchmen: it opposed the fantastic religiosity of the evangelicals and was the name of a prevalent philosophical school which preferred long-tested empirical views to the speculations of scholastic metaphysics and Calvinist theology. The ballad was very well known by the time of Stirling's ordination, so the new minister would inevitably be judged with reference to it. And, since its tone is frankly ribald, a less appropriate introduction for a convinced and sincere Christian it is hard to imagine. The poet, of course, was Robert Burns. Its existence meant that any appointment to the second charge of the Laigh Kirk would take place against a blaze of local publicity and gossip.

To people who are not Scottish, it is difficult to explain just what a freight of values and prejudices this introduction carried. By 1816, Burns was already a national hero; he still is. But in the aftermath of the French Revolution and the war against Napoleon, his fame or notoriety knew no bounds. Remember, the level of literacy among the Scottish common people was probably nowhere excelled, and their national poet was also their romantic hero. As a known moderate, Stirling needed serious force of character if he were to occupy the position and be accepted by the congregation. That he did have such force of character is shown by the fact that a week later he lodged the application to patent his heat exchanger and engine.

CHAPTER 2

Stirling's Engines

Stirling's patent application is in two parts. The first, and by far the larger part, is about his heat exchanger. He describes how the hot gases coming out of a furnace may be made to flow past the incoming air and so transfer much of their heat to it. This is a very important invention which was to be used in making the steel which would be so necessary for developments in engineering later in the century. And, ironically, it would make the steels which were needed for Stirling's engine to work acceptably. But it is with the secondary part of the invention that we are concerned: an engine which, Stirling says, may be powered by the waste heat, by means of a modification of his heat exchange invention. And note: by the heat alone. No steam, no oil or anything of the sort. This engine was remarkable. It has often been described as revolutionary – but like many revolutions, it happened a bit too early. About two hundred years too early.

At its most basic, the engine works by using fire to expand air and then cooling it so that it contracts. The idea, in one guise or another, had been around in rudimentary form for a long time – a very long time, depending on whose account you believe. Until Stirling, none of the devices really worked: some of them more-or-less worked some of the time; none of them worked properly all of the time. But an engine was built, probably according to the specifications in Stirling's 1816 patent, which in 1818 was put to service pumping water out of a quarry near Kilmarnock. By all accounts it did its job well, and continued to do so for three years until some fool allowed it to overheat and it went on fire. But three years' pumping is proper work by anyone's standard, so the machine must be taken seriously.

Stirling's machine was a major innovation: in its conceptualisation, in its design and in its execution. The closed-cycle air engine was quite new. Both Sir George Cayley and John Ericsson (see Chapters 3 and 4) were working toward the same end. Cayley never got there; and though Ericsson did, it took him a long time, and he ended up copying Stirling's engine half a century after it had been invented.

The design in the 1816 application for patent is a surprisingly sophisticated one, not at all the sort of thing you would expect an amateur to devise. It embodies many elements which are plainly the result of much thought and experiment. It is well in advance, in design terms, of two models which Stirling is said to have given to Edinburgh and Glasgow universities. It is frankly not credible that either design or execution was the work of a student of divinity or a newly appointed junior clergyman working on his own. As regards execution, the engine required castings of a size which Kilmarnock would have been hard-put to produce. Then their machining and assembly would have required plant, machinery and muscle as well as a lot of people skilled in using all three. We know nothing of them, so we are undoubtedly lacking much of the backstory.

There are really two different inventions in the one machine. It came to be called the regenerative air engine: it is an engine which works by the alternate expansion and contraction of air, and it embodies a regenerator (Stirling called it an 'economiser'). The expansion and contraction of the air in the engine is what makes its wheels turn; the regenerator is what makes it go *well*. It will run without the regenerator, but badly: you get some bangs for your buck, but not nearly as many as you get if there is a regenerator. The regenerator is the bit which is hard to understand.

Three engines are known to have been built to Stirling's designs in the years around 1816: the two models mentioned above and one full-size, the latter probably being that described in the 1816 patent. Almost nothing is known for sure about the origin of any of the engines: no drawings of the models are known to exist (though the models themselves do) and no trace of the engine built to the 1816 drawings survives (though we have the drawings).

The principle on which the engines work is the same in all three cases. Given the air engine has pistons moving in cylinders, it does not appear entirely unfamiliar to anyone who knows a bit about how internal combustion engines work.

Briefly, in an internal combustion engine, air plus an inflammable gas or vapour are detonated inside a cylinder fitted with a moveable piston which is connected to a shaft. The explosion causes the piston to move and the motion of the connecting rod turns the shaft. There are two main sorts of internal combustion engine: petrol and diesel. In petrol engines, the explosive gas is liquid petroleum, which is gasified by a carburettor or injector; it is ignited by a spark plug. A diesel engine uses a different sort of oil and the explosion is caused by the rise in temperature resulting from the compression of the fuel/air mixture. Diesel engines have to be built heavier than petrol, to cope with the stress of the higher compression. They are all called *internal* combustion engines, to contrast them with the like of steam engines, where the combustion of the fuel is *external* to the engine.

But even for someone who knows how internal combustion engines work, the Stirling engine is not easy to understand, and for that reason I do not propose to begin this chapter with the description in the 1816 patent specification, but with the two models, which are radically different in design from the 1816 engine and much simpler.

The Edinburgh Engine

There is a rather grand photo of the Edinburgh engine (plate 4): it's not nearly so grand in the flesh, so to speak, being quite small, and its frame made of wood. (It looks fancy because the National Museum of Scotland, to which it was given by Edinburgh University some years ago, evidently thought it was important enough to have smart photos taken of it.) As you can see from the photograph, it's the same layout as the engine which I bought on the Portobello Road, only arranged vertically rather than horizontally.

There are two cylinders set in the frame, which also supports the shaft which, by cranks, links the two pistons and bears the flywheel. A pipe goes from one cylinder to the other, so air in the one is connected to the air in the other. From the crank at either end of the shaft, a rod goes into one of the cylinders and at the end of each rod there is a piston. The rod to the larger cylinder goes through a stuffing box so that the air inside the cylinder is contained. The piston in the smaller of the cylinders is a gastight fit; the piston in the bigger one fits only loosely so that when it moves up and down the air is shifted from one end to the other, being squeezed past the piston. This piston is almost as long as the cylinder in which it moves, and at each stroke one end comes close to an end of the cylinder, so the result of the motion is to send almost all the air first to one end and then the other. This piston is accordingly called the displacer, while the other is referred to as the power piston, or simply the piston. The cutaway diagram of the 1816 patent application on p. 42 may make things plainer.

Heat is applied to the bottom of the larger cylinder at a point when the air is at that end. The cylinder being made of copper, the heat is quickly conducted through its bottom and transferred to the air inside the cylinder. When the air inside the large cylinder is heated, it expands and, because it can't escape, its pressure rises. Because all the air in the engine is connected (remember the displacer doesn't fit the cylinder and there is a pipe connecting both cylinders), the air pressure throughout the engine rises to become greater than the pressure of the outside atmosphere. That means that the pressure inside the piston is greater than that outside, so the piston moves as far as it can. When it moves, it causes the shaft and the flywheel to move, too – and the other end of the shaft is connected to the piston rod at such an angle that the motion pushes the displacer to the bottom of its cylinder. When that happens, the air is shifted to the top of the cylinder, where it loses heat to the atmosphere; the pressure inside drops again and the momentum of the flywheel takes us back to the starting point.

That is how the Stirling engine works – at least it is part of the story. Heat flows through the engine from the hot end to the cool end; some of the heat (about two-thirds) is lost at the cool end, but a significant part of it – about a third – is converted into mechanical work. It is a machine of great ingenuity and – as we shall see later – subtlety and at the time it was invented it was perfectly unique because *it worked*. The idea of using the expansion of air to produce mechanical effect wasn't new, but this was the real McCoy. (George Cayley, among others, had tried it but hadn't got it to work properly.)

It will come as no surprise to the reader to find that there is a frustrating lack of hard evidence about this model engine: why it was given, when and to whom. But if it was presented in the early 1820s, as is generally believed to be the case, the likelihood is that it was given to the professor of natural philosophy – who at that time was our friend John Leslie. There is a strong probability that Stirling and Leslie became acquainted when Stirling was at Edinburgh and that the acquaintance developed into friendship and collaboration, and led to the production of Stirling's first engine. This would explain why, after four years of divinity in Glasgow, Stirling spent another year in Edinburgh before seeking ordination as a minister. (Leslie became professor of natural philosophy on the death of John Playfair in 1819 and held the job until his death in 1832.) The presumption of a close relationship is further supported by the knowledge that John Leslie's nephew, James Leslie, was one of those who in 1834 proposed James Stirling for membership of the Institution of Civil Engineers and in 1839 married the sister of Stirling's wife Susan. It is a small country, and its population was a lot smaller then.

In both Edinburgh and Glasgow the professors of natural philosophy illustrated their lectures by means of models which exhibited the processes under discussion. It is known that Leslie had a large collection of such equipment, either made by himself or by a professional instrument-maker. It's a reasonable presumption that the purpose of the gift was to add to this equipment – which if true implies that the expansion of air and its consequences formed a significant part of Leslie's lectures – and that his relationship with Stirling had become a collaborative one. It is also my guess that during that lost year at Edinburgh in 1815 Leslie and/or his instrument-maker collaborated with Stirling in producing the model; and perhaps in making a single-cylinder machine, now lost, on which the patent application was based. We know that Leslie used Stirling's engines in his demonstrations in the 1820s.

The Glasgow Engine

The general arrangement of the Glasgow engine is the same as that of the Edinburgh one, with an exception. The story goes that the Edinburgh engine, when supplied with heat to the bottom of the displacer cylinder, would run merrily – for a time. But after a bit, it would slow down and eventually stop. This is because these engines need to be supplied with heat *and to have it taken away* – the removal is as important as the supply. When the Edinburgh engine was first heated, one end of the displacer cylinder became hot and the other stayed cold, but after a little while heat was conducted from one to the other by way of the casing and the displacer: the cold end became hot too, so the flow of heat ceased, or was greatly diminished. (If you happen to own one of the little Chinese-made Stirling engines which run on the heat from a teacup, and find it soon stops, try putting a piece of ice on the top plate. The engine will start again because the heat is being removed.)

Stirling appears to have understood this, for the Glasgow engine is supplied with a jacket which surrounds the cold end and the jacket has pipes through which cold water can be supplied to take away the surplus heat. It will run well as long as the flow of water is maintained. Just when Stirling came to such an understanding is unclear.

It seems likely that Stirling was close enough to the staff of Glasgow college to present them with an engine for their practical demonstrations of the motive power of heat. Alas, it can't be said that the Glasgow people were properly appreciative, for thirty years later, when William Thomson, who as Lord Kelvin was to become one of the titans of nineteenth-century engineering, took the chair of natural philosophy at Glasgow, he found Stirling's engine languishing unloved in what he referred to as the department's 'augean stables' of experimental instruments. Thomson at least esteemed the machine, had it cleaned up and lubricated, and used it as an important teaching aid. We shall return to Thomson later.

The 1816 Engine

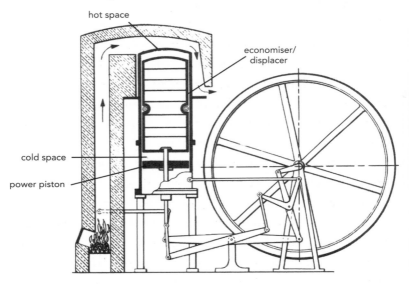

hot space

economiser/
displacer

cold space

power piston

Diagram of the 1816 engine from the patent application.

We can now turn to the engine which features in the 1816 patent application and is probably the engine as it was built for pumping out a quarry in 1818. As can be seen in the illustration, it looks very different from the models, but the difference is in design, not principle. Instead of having two cylinders, the engine has only one. The power piston is in the lower end and the displacer takes up most of the room in the upper part of the cylinder. The hot end of the cylinder is the top end and the bottom end is cool. The air inside is shifted from top to bottom and back again by the displacer and the power piston is driven down by the increased pressure of the heated air. No provision for removal of heat is shown, probably to simplify the illustration, but there was undoubtedly some such when the engine was built. The rod leading to the displacer passes through the piston, so that piston and displacer can move independently of each other.

Since the patent was to be for a practical machine, it is shown as having a source of heat: on the left side is a coal fire with a flue, which takes the hot gases to the top of the cylinder and then leads to a chimney which vents them to the air.

The flywheel is shown in the illustration, as are the levers which connect the piston rods to the crankshaft. (The levers look very complicated but are pretty simple in operation: steam engine builders had long been familiar with such linkages, whose business is to determine the timing of the pistons in relation to the revolutions of the shaft and flywheel.) Apart from all that, the big difference between this and the model engines is in the depiction of the displacer – and that brings us to a matter which is at the core of Stirling's engine.

The displacer which Stirling shows is evidently intended to do more than just displace, unlike the displacers in both the model engines. He tells us in the application for the Scottish patent that

> it is a hollow cylinder ... as thin as possible, made of sheet iron, covered with thin plates of polished brass or silver, to prevent the waster

from radiation, and divided into compartments by plates, for the same purpose. It is shut on all sides, and air-tight, kept at a small distance from the outer cylinder by wheels or any similar contrivance, and furnished with a rod . . . working through a stuffing box in the centre of the cylinder by means of which it is moved up and down . . .

The interval between the cylinder and plunger is partially filled with wires wound round the latter, and kept at a small distance from it and from one another by wires laid along it at right angles to the former, *in order to heat and cool the air more completely.* [My italics.]

It will be apparent to the discerning reader that there is a disparity between the elaborate description of how the displacer is constructed and the rather terse description of what it does. Nor is it apparent that its construction will facilitate its function; indeed, there is reason to suspect that it may be counter-productive. If the conversion of the heat of the fire into the mechanical action of the fly-wheel depends on efficient movement of hot air through the cylinder, all those wires would seem to be more of an obstruction than a facilitation. And why would all that stuff inside the air-tight displacer help to do what he says?

A little later in the same document, Stirling says that when the plunger is pushed down from the hot end, the cold air which is squeezed past it on its way up from the cold end is heated – and the reverse on the way down.

The brilliance of the invention is not in dispute. The best part of two centuries would pass before the economiser was properly understood. The big question is: how well did Stirling understand what he was doing? And do *we* understand what Stirling did? His design is dependent on two sorts of knowledge: practical experience of making engines which work, and a theoretical understanding of what heat is and how it is capable of producing motion. It is hard to see how Stirling could have gained much of the first in his short career up to 1816; his equipment as regards the second we shall now consider.

He knows about steam: how boiling water will produce steam at pressure if prevented from escaping. (By 1816 high-pressure steam engines were well known, if not yet common in Scotland.) He knows such pressure can force a piston to move and do work. He knows that air and other gases will expand if heated and similarly exert pressure. If asked what heat consists of and how this happens, he will tell you that it is due to the presence in hot things of Caloric.

In Stirling's day there were two theories competing to explain what heat was and why it acted as it did: the Caloric and the Kinetic theories. The latter was the closer to modern ideas, but the Caloric theory was more widely held. Proposed by the chemist Antoine Lavoisier about thirty years earlier, the theory held that heat was in essence a subtle, invisible, weightless, self-repelling fluid which would naturally flow from a hot body to a cold one. The more Caloric a body contained, the hotter it would be. It was a pretty good theory, even if it was later superseded, for besides being easy to understand (the obvious analogy was with the behaviour of water), it made sense of most of the main ways in which heat behaved. It could more or less explain radiation, convection, phase change and the expansion of gases. It also enabled the polymathic Pierre-Simon Laplace to correct Newton's calculation of the speed of sound, so clearly it was not a theory to be sniffed at. But even by Stirling's time it was being seriously questioned.

Fifty years before, Joseph Black had conducted a series of investigations which would become classics of experimental method. He had demonstrated that the amount of heat which a body held depended not only on its temperature, but also on its weight and on the material of which it was made. Different materials had differing abilities to hold heat, which Black would describe as their specific heats. The Caloric theorists found this difficult but not impossible to counter: they would say that, obviously, different materials would hold differing quantities of Caloric – which on the analogy of water seemed plausible, if less than persuasive. A more serious challenge to the Caloric theory was to come from an altogether different and unexpected source.

Benjamin Thompson, Count Rumford, was an American-born scientist, inventor and self-publicist – and a hustler of genius. (We shall meet another in a later chapter.) A loyalist in the War of Independence, as a young man he was welcomed in London, where he quickly became prominent in scientific circles. Born in 1753, by the age of thirty-one he would be knighted by King George III and at thirty-eight made a Count of the Holy Roman Empire by the King of Bavaria, whose entire army he had reorganised. (It was in need of it.) The Rumford was Rumford in New Hampshire, not Essex.

Foremost among his many and varied interests was the study of heat. By age twenty-eight he had published in the *Philosophical Transactions*, the Royal Society journal, on the propellant force of gunpowder. In due course he was made a member. Thompson spent eleven years in Bavaria; while there he conducted experiments which demonstrated the heating effect of boring a cannon using a blunt boring tool. Publishing his results in 1798, he demonstrated that the heat generated would continue to be produced as long as the boring persisted – a conclusion incompatible with the presence of a limited quantity of Caloric. He also demonstrated that no physical alteration in the material of which the cannon was made had taken place. What had happened was that a great deal of motion had been communicated to the cannon and that motion had shown itself as heat. There were no immediate consequences arising from Rumford's experiments but by the middle of the nineteenth century they would lead to J. P. Joule's mechanical equivalent of heat and eventually be absorbed in a synoptic understanding of thermodynamics and kinetic energy. Joule – brewer, physicist and mathematician – would collaborate with William Thomson to effect a revolution in the understanding of the relationship between heat and other forms of energy.

It may be worth saying at this point that, in general, scientific theories aren't simply true, in the sense that we usually give to the word 'true'. They are always to some extent provisional, and capable of being abandoned or adapted if a better one comes along. In this, they are unlike the grand truths of religion or the small truths of common sense. You can sometimes listen to a scientist talking about a theory for quite a long time without hearing him or her mention truth: he or she may describe a theory as useful, or elegant, or persuasive – none of which terms he or she would use of the proposition that God is in His Heaven or that the cat is on the mat. It is another characteristic of a well-developed theory that adherents will not abandon it on account of a single countervailing observation. A much more likely course to take is to seek to articulate the theory in such a way that it can accommodate the contrary evidence. And, of course, one can choose simply to ignore it. (This is a lot more common than the non-scientist might imagine. It is almost certainly the course Stirling must have taken, for he would have known about Rumford's experiments.)

Stirling believed in the Caloric theory: not in the way that he believed in God (he was, after all, an ordained minister) but in the sense that it explained some of what was going on some of the time and because he could see nothing better on the horizon. His problem was that the flow of Caloric could explain some of what was happening in his engine, but only vaguely: he was struggling to say just why the thing ran at all, and unable to say why his great invention, the regenerator, made it run so much better. For the latter, at least, he really can't be blamed: here is Allan Organ, a leading researcher in the field of the Stirling engine, in 2016: 'The 20th-century mathematicians *never* solved the regenerator problem.'[1] This in a book,

---

1   Allan J. Organ, *A Re-appraisal of the Stirling Engine*, p.9.

the second half of which is computer code, which I expect you must use if you are to understand the first half, which is dense thermodynamic theory in the form of advanced mathematical reasoning. Only in the last year or two does the thing seem to have approached a resolution: the engineers, having created what Win Rampen (Professor of Energy Storage at Edinburgh University) calls 'clever computational fluid dynamic solvers', have made it possible to devise analytical tools to 'create models which move millisecond by millisecond, micron by micron, through a flow field to really model what is happening in terms of flow and heat transfer'.[2] Win describes how the regenerator picks up heat at a point in the cycle when its expansive presence would be counterproductive, retains it *in a non-expansive form* and then releases it just when required to contribute to the expansion of the gas in the cylinder which drives the piston and does the work of the engine.

We can find evidence of Stirling's difficulty in understanding his engine by looking at the first three machines: the two models and the 1816 patent. The Edinburgh model, as we have already remarked, had no provision for removal of heat, which rather suggests that Stirling didn't know that the heat had to be removed as well as supplied. The Glasgow engine had a water jacket which put that right. There are two possibilities here: Stirling may have learned from the Edinburgh engine's tendency to slow down and stop as it got hot – or the water jacket may have been added by William Thomson when he fettled the machine for use in his classroom. This takes us back to

---

2   Win Rampen, personal communication.

my earlier assumption that in fact the model engines were made before 1816, when Stirling was in Edinburgh, and that the engine in the patent was a development based on the experience gained using the models, probably with the assistance of John Leslie and others. That is certainly a more likely scenario than the postulate of a lone genius who, out of his fertile brain alone, produced a well-developed, fairly sophisticated design for an engine. The model engines are exactly the sort of things an inventor would build in order to test a hypothesis; the 1816 machine is the sort of thing he might patent thereafter. (In saying this, I do not intend in the least to disparage Stirling or to diminish the originality of his ideas: merely to point out the process of development which would be likely to lead to a well-thought-out patent like that of 1816.

The study of the first three engines suggests that cooling was adopted reluctantly: the 1816 patent application, while referring to it, says little about it: 'I apply a stream of cold air or water to the coldest part of the engine to carry off said waste heat.' It does appear in subsequent engines, but never prominently. The reference to 'waste heat' here is perhaps significant, for there is reason to believe that Stirling thought that in an ideal engine – one in which there were no incidental losses of heat due to conduction, radiation and friction – the whole of the heat applied would be converted into motion. We have one priceless piece of hard information on this: in April of 1848, Robert Stirling had a meeting with James and William Thomson in Glasgow. (William Thomson was one of the greatest engineers and scientists of the nineteenth century; he made Glasgow University a leader in the field and was ennobled as Lord Kelvin by a grateful government.) James kept a manuscript note of the interview: it is in the archives of Queens University Belfast. It is worth quoting here. The punctuation is exactly as per the manuscript.

JAMES THOMSON

Note of Interview with Robert Stirling

April 28 1848

I have just seen the Rev Dr Stirling of Galston, he called here and I had a good deal of conversation with him regarding the air engine. At the commencement, in presence of William and Mansell, and afterwards in the presence of my father, I told him particularly not to tell me anything that he did not regard as entirely public in case I had some ideas on the subject myself. He mentioned that he is at present going on with some improvements in his air engines. Going away, he assured me that he had told me nothing that is not publicly known. I found that, as I had previously thought, he does not understand his own engine; not knowing at all the way in which the heat is expended in generating the work. He said he had been greatly perplexed for a long time about the changes of temperature of the air produced by changes of pressure and that at one time this had rather alarmed him in regard to the perfection of the engine as it appeared that the respirator would not even <u>theoretically</u> give back all the heat to the air; but that now he is inclined to think that a "sort of average is struck" or a compensation is made by which all the heat is really given back if the air passages be small enough the metal perfectly absorbent and non-conducting. I told him that some transference of heat from the furnace to the water by means of the changes of temperature of the air is essential to the action of the engine; that otherwise it would be theoretically perpetual motion. He replied that there are plenty of theoretically perpetual motions if we leave friction resistances etc out of consideration. I said that there are these, but not perpetual sources of power. After some consideration he replied that perhaps what I said was correct and that he had never thought particularly on the difference between a perpetual motion and a perpetual source of power. In pointing out to me what he supposed to be the action of the air in the respirator and so endeavouring to prove that the respirator does really return all the heat to the air, and so that the machine is theoretically a perpetual source of power, he had no idea that the air ought to tend to be cooled by expansion at one part of the stroke so as to take in heat from the fire, and that at another it ought to tend to become heated by compression so as to make it give out heat at the lower temperature; but he was strongly impressed with the supposition that the fire is useful merely to give a small supplement to the heat returned by the respirator so as to make up for incidental losses due to practical

imperfections of the apparatus, such as conduction of heat, incomplete absorption, of work, etc,
not that the removal of some heat from the fire is
essentially connected with the development.

From this it appears that Stirling thought that his regenerator, having acquired Caloric from the furnace, used that
Caloric to produce the power stroke of the piston – and
could go on doing so indefinitely, provided the Caloric
wasn't dissipated by accidental losses such as conduction
and radiation. He believed that the heat which flowed out
of the engine was simply incidental waste and in no way
intrinsic to the operation of his engine. And he thought
that if he got the engine design just right, he would be
able to eliminate all the incidental losses – hence his cava-
lier attitude to the extraction of waste heat. And this was
more than thirty years after he had lodged his application
for his first patent.

He would not have been happy to learn later in the
century of an observation which came to have the grand
title of the Kelvin–Planck Statement of the Second Law
of Thermodynamics. This says, 'It is impossible to construct a device which operates on a cycle and produces no
other effect than the transfer of heat from a single body
in order to produce work.' What that means is that in a
heat engine, you get work from some of the heat you put
in, only if most of it goes straight through.

The Second Law of Thermodynamics sounds like
serious, established science – and around the date of
the interview a loosely connected set of observations
about heat and its effects was coming to be regarded as

a respectable branch of the physical sciences. The first occurrence of the word 'thermodynamics' in English publishing was in a paper read by William Thomson to the Royal Society of Edinburgh in 1849, in which he refers to 'a perfect thermodynamic engine'.[3]

The foundations which had been laid by Black, with his notions of heat capacity and latent heat, would be brought into a theoretical framework by Sadi Carnot and given system by Rudolf Clausius and others to coalesce ideas of heat, mechanical motion, power, work and engine efficiency. J. P. Joule in Manchester, using some of the simplest instruments, would produce a far-reaching effect by demonstrating beyond reasonable doubt that work and heat were exchangeable aspects – of what would come to be called energy. (Joule set up an apparatus in which a weight, attached to a string, the other end of which was wound round a shaft, caused the shaft to turn. If the shaft was attached to a set of vanes immersed in water, the weight would cause the water to be stirred. Joule measured the rise in temperature caused by the vanes' stirring the water over a period.)

James MacQuorn Rankine, in 1859, would publish the first textbook of thermodynamics. Rankine, Thomson and Clausius in the 1850s would set the seal of scientific respectability on the discipline by enunciating the Laws of Thermodynamics. Of the latter, the first two are germane to our thesis. The First Law, the Law of the Conservation of Energy, which states that energy cannot be created or destroyed but only changed from one form

---

3 *Transactions of Royal Society of Edinburgh*, XVI, p.545.

to another, seems obvious to us now – but was not so then. The Second Law has many forms of expression, for it has ramifications in so many fields that it has come to be thought ubiquitous – as, in a real sense, it can be said to be. In the context with which we are dealing here, it can be stated as the simple proposition that heat will flow from a hot thing to a cold one and not the other way round unless some external compulsion is applied. When allied to the atomic and molecular theory of matter, the theory would give rise at the hands of Rudolf Clausius, Josiah Gibbs, James Clerk Maxwell, Max Planck and Ludwig Boltzmann to statistical thermodynamics, a discipline which would allow the theory to permeate much of twentieth-century science.

Thermodynamics would become heavily mathematical and, like much mathematical science, could be accused of forgetting that it began with simple observations of uniformities in nature. The good and great James Clerk Maxwell apparently thought so, for he went to the trouble of devising a mythic scenario which would remind thermodynamicists of the theory's lowly origins. It consists of a closed box in which the molecules of a gas are zipping randomly about – as molecules of gases generally do. The box is divided in two by a partition which is impermeable to the molecules, save for a hole which is covered by a door. The door is worked by a tiny creature who can see molecules coming and is disposed to open the door so as to allow some of the molecules through the hole. It allows fast-moving molecules to enter the left-hand side and slow-moving molecules to enter the right, so that in time the left side will become hotter than the right.

This, if it can be allowed to be possible, will violate the Second Law. Thomson called it Maxwell's Demon and allowed that explaining why it is impossible presents a serious problem for adherents of the prevailing thermodynamic theory. Maxwell's Demon has had long legs: arguments about it have spanned part of the nineteenth century and the whole of the twentieth, taking it into fields which could not have been imagined by its deviser – though many of the more arcane fields of science and technology today owe their existence to the same Maxwell. It is, of course, possible that it was just a joke on Maxwell's part, or a veiled warning that at root the whole discipline was no more than a statistical elaboration of observation.

By the 1850s it had become apparent that the Caloric theory had been left far behind. Quite apart from the impossibility of showing that Caloric exists, the theory was incompatible with the first two Laws of Thermodynamics – and the latter had become immersed in such a vast web of related theories, all of whose predictions agreed exactly with observational evidence, that few thinkers exposed to the weight of contemporary ideas could any longer adhere to it. Robert Stirling had evidently failed to keep up with modern thinking in Galston. The cholera may have had something to do with it, of course.

•

Stirling, together with his younger brother James, would patent more engines, each of them conceived on a much larger scale than the first. None followed the pattern of the 1816 engine, but were closer in design to the two models, having separate cylinders for piston and displacer. All featured regenerators, but the variations in the construction and materials of these suggest that the Caloric theory got in the way of empirical improvement. Their regenerators are constructed using non-conducting materials such as glass and are filled with improbable stuff like brick dust. The Stirling brothers evidently thought that conduction of heat was a defect – whereas it was what made the regenerator work. And they took steps to prevent radiation, to stop the outflow of Caloric.

In 1827 James Stirling converted a steam engine to Stirling operation with a 26-inch-diameter cylinder for Girdwood & Co. of Glasgow. It ran, but failed to meet expectations of power and fuel economy and was scrapped. This evidently gave rise to thought about why the power output was so low, and another engine was projected and patented in 1827 in the names of Robert and James Stirling.

Engraving of 1827. Conversion of a steam engine.

The 1827 engine was based on a steam beam engine. It addressed some of the problems of the previous engine by having two displacer cylinders feeding a double-acting power piston – and by using air at high pressure, the latter provided by an air pump driven by the engine. It does not appear that an engine was actually built to the 1827 design, though in both the double-acting piston and the high-pressure air it prefigured two of the most important aspects of engines one hundred and fifty years later. James is reported to have been the driver of this patent: no doubt (the still young) Robert was fully occupied by his new parish of Galston, a mining and industrial town a few miles from Kilmarnock.

In 1840, the Stirlings patented yet another version of their engine. This time they departed from the beam engine format and adopted a horizontal power cylinder driving a shaft directly. Two displacer cylinders drove the power cylinder. Air compressed to 150 psi was used as the working gas. The main departure was in the design of the regenerator and in the materials of which it was constructed. The regenerator sat outside the main cylinder, whose piston fitted closely enough for its action to force air back and forward through the separate regenerator.

James became engineer and manager of the Dundee Foundry in 1837. This was a large concern whose operations required a lot of mechanical muscle. In those days, factories would have a single source of power, usually a steam engine, which drove a rotating shaft which, in bearings, ran the length of the principal workplace just beneath the roof. From this shaft, individual machines were driven by means of leather belts which could be

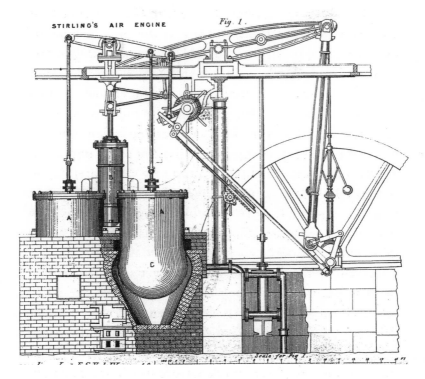

STIRLING'S AIR ENGINE   *Fig. 1.*

Engraving of the 1840 version of the engine.

connected or disconnected by shifting them from a fast pulley – one keyed immovably to the shaft – to a loose pulley which sat over the shaft.

Providing the power for a factory such as the Dundee Foundry cost a lot in coal to drive the steam engine. The Stirling engine was thought (by the Stirlings at least) to be a more economical source of power and in 1842 James converted the steam engine to work as an air engine, presumably along the lines of the 1840 patent, utilising much of the machinery of the steam engine. The steam engine had been rated at 30 horsepower, so the new engine was evidently a sizeable machine. According to the *Dundee Advertiser*, a local newspaper, the savings in coal were expected to run to £2,500 or £3,000 per annum, a very substantial sum in 1842.

It appears, though, that the engine's output did not come up to expectation, for the following year another engine was built which was rated at 45 horsepower. This ran satisfactorily for nearly three years until one of its displacer cylinders cracked. It was repaired, but a few months later the other cylinder went. There appears to have been yet another failure, whereon the owners of the foundry cut their losses by replacing the Stirling engine with a steam engine.

Sundry other people at the time were experimenting with air engines and several were built. Few proved satisfactory and none lasted long. The only material of which the displacer cylinder could be made at the time was cast iron, which was liable to crack because of differential expansion. The Stirlings addressed this by using furnace gases rather than direct heat, which heated the cylinder more uniformly, but the real problem was with the materials and it wasn't until later in the nineteenth century that steel cylinders could be made which resisted the tendency.

—

The Thomson interview is a rare insight into the thought processes of both Stirling and, presumably, his brother. It is astonishing that someone who could produce a major innovation could, after thinking about it for thirty-two years, display so little insight into how the thing worked. That he could admit to not having thought about the difference between a perpetual motion machine and a perpetual source of power beggars the imagination – for the whole point of his engine was to be a *first mover*: a source of power for mines and factories. That's what all engines were about in the early nineteenth century.

If Stirling had been an unreflective engineer who happened to hit it lucky with a single good idea (there are plenty of those), his behaviour might be comprehensible. But he was not. We know from his sermons that he was at ease dealing with abstruse and recondite concepts. (They don't get much more abstruse and recondite than protestant Christian theology: try reading one of Robert's sermons.) So why would he be unable, over a period of thirty years, during which he was frequently engaged with his engine, to think his way through the conceptual basis of its operation? (I'm not talking about the regenerator here – that's *really hard* – but about the relatively simple question of why the thing goes at all.) The only explanation which presents itself is that he was confounded by his adherence to Caloric theory. Having ignored or otherwise discounted Rumford's awkward data, he was stuck with a partial understanding which blinded him to evidence arising from his own researches. For this he ought not to be blamed: it is a commonplace of the history of science. Lots of people had done the same before him – and many still do today, one suspects.

So we have an engine invented and developed by a man who didn't really understand how it worked. But there is worse: it was probably only a matter of time before some-body else invented the closed-cycle air engine, for the idea was in the air, and such things tend to come in clusters. But this same man invented the regenerator, which was in a different league altogether. Where did he get the idea for that from? And how did it come to be so fully devel-oped in his 1816 patent application – especially given that his understanding of the forces and processes was at best only partial? There can be no doubt that it worked, and worked well, right from the beginning. That was the strength of the engine: its other parts were mostly weak-nesses. In his descriptions of its structure and functioning, it is clear that Stirling adhered to the Caloric theory of heat: he thought that the regenerator, once charged with Caloric, would go on producing force indefinitely.

The best explanation of the working of the regener-ator which I have come across (and I have to admit to being an amateur and an outsider in these arcane matters, so I say it with some trepidation) comes from Professor Win Rampen, who explains, using the diagrams which his computer produces of the fluid dynamics of the movement of heat around the regenerator. It's about how heat flows into and out of the regenerator, and at what point in the cycle. Win points out that in a Stirling cycle without a regenerator a lot of heat is present in the air at a time when its presence is counter-productive. Think of the two-cylinder form of the engine: when the air is at the hot end of the cylinder, the heat flows into it and, by con-duction and convection, all the air in the engine expands and causes the pressure to rise. The rise in pressure causes the piston to move out; the displacer moves in and squirts the air to the cold end where heat is lost, while the piston comes back in. But at the point at which the piston begins to move in, little heat has been lost. In consequence, the pressure of the air is still higher than that outside, so the piston meets resistance, to overcome which it must use some of the power of the engine.

What the regenerator does is absorb heat by conduction and radiation from the hot end of the cylinder. It stores that heat when it isn't wanted; and it stores it without expanding the air, and releases it only when expansion of the air will be advantageous to the working of the engine and hence to the creation of force on the outwardly moving piston. There is a lot more, but that's the essence of it as far as I understand. Besides explaining why the regenerator improves the efficiency of the engine, it explains why the engine will work without a regenerator, but not well.

This is clever stuff: you have to be smart to understand it; smarter by orders of magnitude to invent it in the first place – unless your invention is by mere accident. That, of course, is the big question: *was* it mere accident? Or did the tutelary deity of practical engineers (the one with the dirty fingernails) inspire it? Did he breathe it into Robert Stirling – a person who didn't believe in tutelary deities? Or, as is more likely, did Stirling understand on some level which we do not know about? Some skill or some feeling for how things work? Our comprehension of human intuition lags well behind our understanding of many other things: may it be that at some point we shall come to an apprehension that some people simply see how things are, or ought to be? That is, after all, the basis of hypotheses, which are the foundation of scientific understanding – even when they *are* wrong.

———

The Stirling engine, though it undoubtedly worked, was not a commercial success for more than half a century after it was invented. Its enemies were twofold: they were thermally induced stress in cylinders, and other engines (see later chapters). The Stirlings were not the only people to build heat engines and others soon would make closed-cycle variants of their engine. None survived: the cylinders always gave out – and replacing cylinders was an expensive business. In addition to the cost of making new ones, there was the cost of the downtime of the engine. Remember, it was to be used as a first mover, which meant that a great many operations which used it as their only power source would stop when it did. Steam engine boilers were a lot easier: they had to be heated only up to a fraction over 100 degrees Centigrade to produce steam – against 600 degrees for a Stirling engine – and once the steam was being produced, their temperature would be more or less constant. Stirling engines, on the other hand, had to be raised to red heat for the engines to work. The heating induced expansion stresses in the cast iron of the cylinder base where the heat of the fire was applied, and the stresses would eventually cause the cylinder to fail. Had the Stirling engines consistently produced the savings in fuel which were claimed for them, that might just have been tolerable – but it seems that they did not, or did not do so consistently enough for the savings to outweigh the cost of replacement.

And there was the first appearance of the *other engines* problem. By the mid-nineteenth century the steam engine was on a roll: the development of the technology became self-sustaining, given the vast penumbra of ancillary industries which supported it, and which in turn were supported by it. There was no niche into which the Stirling engine would fit for another generation or two. It would happen, but in a limited way, in situations to which steam was unsuited. And in its turn, the Stirling technology would become redundant because of other engines. But its time would come.

CHAPTER 3

Sir George Cayley

Around the time that young Robert Stirling was studying at Edinburgh University, the engineer Sir George Cayley wrote a letter to science periodical *Nicholson's Journal* in which he described an engine he had invented. Like that which Stirling was later to devise, Cayley's engine worked by expanding air. The letter, from Brompton in Yorkshire, and dated 25 September 1807, begins:

> Sir, I observed in your last Vol P368 that some experiments have lately been made in France upon air, expanded by heat, applied as a first mover for mechanical purposes. This idea, as you justly remark, is by no means new in this country; yet I have not heard that any successful experiments have been made, exclusively upon this principle, in England, though you hint that something promising has been accomplished relative to it.
>
> The subject is of too much importance as the steam engine has hitherto proved too weighty and too cumbrous for most purposes of locomotion; whereas the expansion of air seems calculated to

supply a mover free from these defects. Under
this impression I send you a sketch of an engine I
projected upon this principle several years ago; it
was made on a considerable scale at Newcastle,
though I must confess without success in the
result, which I attributed to the imperfect manner
in which it was executed, the cylinders being
made of sheet copper and so irregular, as not to
be rendered tolerably airtight by any packing of
the piston. I think there can be no doubt that
the scheme is practicable in one way or another;
and I conceive that the form of the engine here
sketched will be the basis of whatever experience
may prove to perfection in the apparatus of the
air engine.

Cayley's machine, as did Stirling's engine, uses air
expanded (and thus pressurised) by heat. There are two
working cylinders of different diameter, one above the
other, and each has a piston which is fixed to a single
piston rod. In a separate cylinder to the right, there is a
supply of coal and a firebox. As the lower piston descends,
it sucks cold air in at the top left, while the air already
in the cylinder is pushed down at the bottom right. The
latter charge of air is pushed up and then down through
the fire, then up again into the upper cylinder. Because
the air has passed through the fire, it expands so that the
pressure of air in the upper cylinder is greater than that
in the lower. That difference is what makes the thing run.
(Of course it needs valves and linkages to make it work,
but the principle was sound and the engine produced

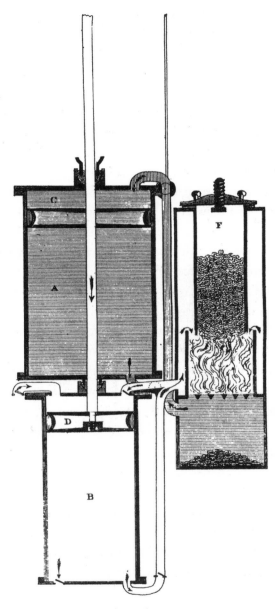

Cayley's first engine.

useable power.) If the expansion of air alone were to provide insufficient force, Cayley suggests additives:

> If, when the engine is well-constructed, the
> expansion of air in keeping up the fire be found
> sufficiently sensible, still, the form of the engine
> is such as to admit either of inflammable gas, and
> of tar and of other inflammable matters, being
> injected, each stroke, upon the fire, so that all
> the heat generated by the united combustion may
> operate without waste . . .[1]

The prospect of the injection of tar, gas, etc. is a trifle
alarming, given the state of the technology. But they
pale beside Cayley's proposal to use gunpowder. In a
later paper he describes a gunpowder engine: a bizarre
machine, a hybrid of gun and crossbow which, happily
for all concerned, was never built.

The mode of operation of Cayley's air engine is, if anything, even harder for the uninitiated to understand than
Stirling's. There are three ways in which air expanded by
heat was to be used to produce power and it may help if
we set them out.

1. Stirling's engine, which we have already described, is a closed-cycle engine. The working
   gas is never exposed directly to the heat source
   and is reused time and again. The heat source
   is external to the working fluid, which means

---

1 *Sir George Cayley's Notebook*, 22 November 1807

heat must be conducted through the end of the displacer cylinder.

Both the other types are open-cycle engines:

2. Open-cycle clean engines, where new working gas is taken in for each cycle, but the heating is external to the working fluid. See Chapter 4.
3. Open-cycle dirty engines, in which the products of combustion mix with each new charge of working gas and are introduced to the power cylinder.[2]

Cayley's engine was of the third type: an open-cycle engine, but a dirty one. From time to time, Cayley would make improvements to his engine which would bring it a modest success, if never the breakthrough which was his objective. The problems with the cylinder would be overcome by using castings, but the difficulty of obtaining a good seal between piston and cylinder was endemic, for the products of combustion – gas, ash, grit, smoke and attendant sulphur compounds – were introduced along with heated air to the cylinder. Piston seals of leather greased with tallow (which is what Cayley used) would never cope with those.

Cayley had an engine built which he installed in the brewhouse of his residence, Brompton Hall, and which ran in an acceptable manner for many years. He seems to

---

2   A. Slaby, 1878, quoted in Finkelstein & Organ, *Air Engines*, p.2.

have had an adequate staff of people whom he had trained as mechanics. It is a fairly safe assumption that the defects of his engine in this case were overcome by the labour and expertise of his staff, who presumably would have had enough time to ensure that it ran well, given that it would have been seen by Cayley as a demonstration piece. (The scale of Cayley's operations may be gauged by the fact that Brompton Hall was of a size large enough to warrant a separate brewhouse with attendant staff.

For Cayley, the engine was only one among many interests, and its invention was to meet a felt need, for Cayley had invented an aeroplane and he needed an engine to push (or pull) it along. (It would not be extravagant to use the definite article here and say that Cayley had invented *the* aeroplane, for the device he had in mind was the basic layout for most aircraft thereafter, of wing mounted on a fuselage with behind it a tailplane with vertical fin for lateral steering and elevators to point the aircraft up or down.)

There is reason to believe that Cayley's basic design for an aircraft was produced as early as 1799, when he was twenty-six years old. The evidence for this is of a rather romantic kind: in 1935 a Mr Smith of Scarborough received a small silver disc from one of the Cayley family. On the disc are engraved the date 1799 and the letters GC. On one side of the disc is a sketch of a monoplane glider aircraft to Cayley's design; on the other, a geometrical symbol representing the principal forces acting on any aircraft: thrust, drag, weight and lift. Mr Smith passed the disc to the Science Museum, where it now rests. There is no other evidence at all, save Cayley's own assertion in his 1809 paper that his first experiment in the matter of heavier-than-air flight was in 1796.

Like almost everyone who considered the possibility of flight by a heavier-than-air machine, Cayley studied bird flight. But unlike most such investigators he concluded that the way to gain lift was not to try to imitate the downward stroke of a bird's wing (it was mechanically just too difficult to contrive, Leonardo notwithstanding). Cayley observed that in flight many birds appeared to stay aloft by exerting little effort, and concluded that, besides the upward force exerted by a downswept wing, the bird relied for its lift on the effect of its forward movement through the air. This, together with his observation that a bird's wing was cambered, was a major breakthrough in understanding flight.

To test his theories, Cayley constructed several gliders in his workshop at Brompton Hall: in 1804 he built one with a wing area of 200 square feet, and in 1808 another of 300 square feet: both flew well. Throughout his life, Cayley would continue to build gliders, while developing his air engine as a means of propelling them through the air. (He had devised a rudimentary propeller.) In 1849, he built a glider which carried a boy as a passenger and in 1853, a triplane machine on which he persuaded his coachman to act as pilot. The latter flew for a distance of 900 feet across Brompton Vale before crashing. It is reported that the coachman was unimpressed by his status as the first human being to fly and complained that his duties as coachman did not oblige him to risk life and limb in aeronautical devices. Had Cayley possessed an internal combustion engine like the Wright Brothers', there can be little doubt that the first powered human flight would have been made in Yorkshire some decades before Kitty Hawk Down.

—

Cayley was born in 1773 with some notable advantages in life. The place of his birth, Paradise, is suggestive, though it must be admitted that it was a house called Paradise in Yorkshire, not the other. Still, young George was to inherit a baronetcy and landed estates sufficient to enable the profligate life of a Regency aristocrat, had he been so inclined. Fortunately for our story he was not so inclined, nor was he inclined to do what most Englishmen in his position would have done, to become a robust, jolly, hunting squire and philistine. From his youth, Cayley was interested in how things worked, and he adopted a remarkably precocious, analytic technique for discovering the same. He was to grow into a quite extraordinary man: an exemplary landlord, an agricultural reformer, a political reformer and, eventually, a Member of Parliament. He was popular, sociable and appears to have been devoid of the snobbishness which was to become such a notable feature of Victorian English society. (As regards the last, he drew on a set of attitudes typical of an older, and in many ways more admirable, society.) But he did share the Victorian conviction of the inevitability of progress, both moral and material, which was to be brought about by discovery and innovation. Innovation was what Cayley was good at: in the course of a long life, he was to invent many things which have become so ubiquitous that their ingenuity is invisible to us. Among the latter are the tension-spoke wheel (think bicycle wheels), the caterpillar tractor, self-righting lifeboats, an internal combustion engine of sorts, seatbelts and railway signals. Fortunately for the biographer, Cayley left copious accounts of his doings: in letters, in notebooks and in papers for learned societies and magazines. Of the latter, by far the most important for our purposes is the letter dated 25 September 1807, to the editor of *Nicholson's Journal*, quoted above.

Until the advent of *Nicholson's Journal*, the only such periodical of any consequence was the Royal Society's publication, the *Philosophical Transactions*. Since membership of the Royal Society was restricted, and its *Transactions* printed only submissions by members or associates, and there were worthy scientists who were not Royal Society members, an opportunity presented itself. This was seized by William Nicholson, who issued his first journal in 1797. Nicholson, himself a member of the Royal Society, was, like many of his contemporaries, something of a polymath. He had papers published in the *Philosophical Transactions* on applied mathematics and electricity, wrote books on physics and chemistry, and with one of his colleagues was the first to decompose water into its constituent elements. He was also a literary critic and invented a printing press.

Even allowing for the fact that science was less specialised in those days, and it was easier to be expert in many fields than it is today, there did seem to be a remarkable number of people around at the beginning of the nineteenth century who would qualify as polymaths. Quite a few of them were friends of either Cayley or Nicholson, or both. *Nicholson's Journal*, being more accessible (for both writers and readers), soon acquired a large readership, with contributors among the good and the great. Of the latter, Humphry Davy, Benjamin Thompson (Count Rumford), Henry Cavendish and Alexander von Humboldt were regular contributors. The science of heat was a common topic. Since *Nicholson's Journal* was widely read, young Robert Stirling would have seen the article and John Leslie would no doubt have remarked upon it in his lectures on the expansive effects of heat.

In November 1809, the *Journal* carried a long essay by Cayley entitled 'On Aerial Navigation'. This was arguably the first properly scientific work on the subject of human flight. Cayley considers first the possibility, which had been reported from Vienna, of a man's having raised himself in the air by mechanical means. (He assumes this means by the use of his own muscles.) Cayley points out that anything of the sort is likely to be of little practical use and says that useful human flight can be obtained only through the employment of a first mover. As we will recall from earlier, the term first mover meant a source of power to engineers of that period, such as a steam engine, a windmill or a waterwheel – there weren't any others to speak of. He does some calculations of a steam engine's weight (engine, water, condenser) and of its power supply (coal) and concludes that the engines of Messrs Boulton & Watt wouldn't be up to the job. He does allude to high-pressure steam engines (such as Trevithick had exhibited some years before) and thinks they may offer possibilities – but evidently doesn't know much about them. (The reason for this is not our concern here: the cause most likely lies in imperfect communication between the very different social milieus of Trevithick and Cayley: the latter an aristocratic gentleman and member of the Royal Society, the former a rough Cornish mine engineer.)

He mentions a machine which, following the French invention of 'igniting inflammable powders in close vessels', has been made to work by the 'inflammation of spirit of tar' but dismisses it because of its undue consumption of fuel. He does think that something of the sort could be done, though, using 'a gas-light apparatus, and by firing the inflammable air generated, with a due proportion of common air, under a piston'. The basic concept of an internal combustion engine is present, but awaiting nearly a century of development.

Cayley goes on to apply his understanding to the principle on which a flying machine might be constructed so as to be capable of carrying a human being. He is confident that 'this noble art will soon be brought home to man's general convenience, and that we shall soon be able to transport ourselves and families, and their goods and chattels, more securely by air than by water, and with a velocity of from 20 to 100 miles per hour'. Plainly, a lunatic visionary.

Perhaps not: after his analysis of how birds fly, Cayley ends by demonstrating that artificial heavier-than-air flight is indeed possible. He describes a toy helicopter which he made in 1796, after a model made by Frenchman Christian de Launoy and his mechanic Bienvenu. It consists of a vertical rod, at either end of which is a cork into which are stuck four goose feathers, at equal intervals. Attached to the bottom cork is a length of whalebone to either end of which is tied the end of a string which goes once round the rod. The rod is fixed to the upper cork but free to rotate relative to the lower. The rod is turned so that the string winds itself around it and the whale-bone bends. When the thing is released, the upper cork will rotate in one direction and the lower in the opposite, and the whole thing will rise into the air. (I made one of these many years ago. It works, but it's difficult to get the whalebone these days, for lack of ladies' corsets, not to mention whales.) The point, however, is a serious one: the toy demonstrates that, given a source of power light enough in proportion to the weight of the craft, a self-sustaining, heavier-than-air vehicle is a possibility.

While Cayley makes it clear that human flight will require an engine with a power/weight ratio far above those of steam engines, he doesn't mention air engines, which is strange, given that just two years earlier, in September 1807, he had written a letter to the editor of the same journal proposing an air engine of his invention as a possible candidate for propelling a (land) vehicle.

It may be worth mentioning that while Cayley was right in supposing that the steam engine would not be the machine which brought about an aeronautical revolution, there were many other people who were convinced that it would. Throughout the nineteenth century, ever-more ingenious high-pressure steam engines would be tried as first movers for aircraft, and inventors would spend time and money in trying to make them fly. In 1848, John Stringfellow would have a small success in persuading a steam-powered model aircraft to fly a few feet – not exactly a triumph, and it did not lead to a full-size craft, as no doubt was intended. But there was enough interest to justify a grand Aeronautical Exhibition in London's Crystal Palace in 1868. The variety of both proto-aircraft and engines was astonishing, though they had in common the single, unfortunate attribute that none of them worked. My favourite from this period is a machine now in the Musée des Arts et Métiers in Paris, which in our family is referred to as the Motorbat. It is an aircraft designed using as a model the wings of a bat: there are long, semi-rigid fingers, between each pair of which is stretched a tight fabric. The motive power is a three-cylinder, triple-expansion steam engine driving a bladed propeller, the steam being provided by an oil-burning boiler. It is a thing of beauty, even if it didn't fly.

———

For the rest of his long life, Cayley would return to his fire engine. For some of the time, he would be assisted by his friend, Sir Goldsworthy Gurney, also something of a polymath. Gurney began adult life as a physician and surgeon but soon turned his attention elsewhere. He is best known for his steam road carriage, which in July 1829 made the journey from London to Bath and back, at an average speed for the return journey of 14 miles per hour, to great astonishment. Gurney would go on to give further demonstrations of the superiority of his machine to horse-drawn vehicles, and to set up a business making and running steam road carriages. Alas, he fell foul of powerful vested interests, whose owners persuaded parliament to allow the imposition of tolls which rendered his carriages prohibitively expensive. A House of Commons select committee enquired into the matter and recommended in his favour that the tolls be removed, but the House of Lords refused to pass the bill and mechanical propulsion disappeared from the turnpikes for another three generations. So much for the Victorian belief in the inevitability of progress.

Cayley was to form a partnership with Gurney in 1837 to develop the air engine, under which they were to share equally in any ensuing profits. It is not clear what, if anything, came of the partnership, though Cayley would continue to improve his engine throughout his life. The most authoritative account of his engines comes probably from a letter he wrote to Charles Babbage when he was seventy-nine years old:

> I made several experimental engines; and the first that succeeded, a one horse-power, made entirely under my own direction here at Brompton, worked for days together without inconvenience from dust or heat. When I got to London my plans were over-ruled by the supposed superiority of London workmanship, and the engine proved more faulty than before – such experimental engines, as you well know! are costly matters; and our funds are exhausted . . . before the engine had a fair chance of being freed from its remaining evils. I am now making one myself again and hope to get quit of the evils of dirt, or over-heated piston.

The content is revealing. Firstly, he is seventy-nine years old and still full of enthusiasm for his engine. More to the point, though, he twice mentions the evils of dust, heat and overheated piston. These were consequences inseparable from the basic design of his engine. Because the expanded air contained the products of combustion, the gas entering the power piston was very dirty indeed. The working of the machine, like that of every other piston engine, depended on a gastight seal between the piston and the cylinder wall – and that is something almost impossible to achieve in the presence of dust, cinders and the corrosive products of the combustion of coal or coke.

Such defects did not render the engine entirely unworkable and so great was the demand for small first movers in the later nineteenth century that large numbers of engines were built along lines similar to those pioneered by Cayley. They would come to be called Furnace Gas Engines; they would be made from massive iron castings and operate in environments where cheap coal was plentifully available, and there was so much smoke and dirt around that a bit more would be no inconvenience. They were so thermally inefficient that ill-fitting pistons were scarcely noticed.

The engine in Cayley's brewhouse.

It has to be admitted that Cayley's engines did not come up to his expectations during his lifetime or thereafter. We can see now that direct-fuelled air engines were never going to make it as power sources for aircraft – and Cayley's design least likely of the three variants to succeed. He put one engine to work in his home brewery, presumably driving a pump or stirring a mashtun, but that is a far cry from powering an aeroplane. The open-cycle dirty arrangement just gave rise to too many difficulties, despite his lifelong persistence. Perhaps if he had been differently situated he might have seen this earlier: all Cayley's engines were made for him by artificers whom he employed. He was an aristocrat and a gentleman, after all, and people of his class didn't get their hands dirty, whereas both Stirling and Ericsson, being middle class, had no such objection and consequently, being closer to their inventions, were more aware of their defects.

—

Cayley's work, especially his 1807 and 1809 contributions to *Nicholson's Journal*, would probably have come to the attention of young Stirling soon after they were published. They would certainly have been known to John Leslie. We can be reasonably confident that if Stirling's idea for an engine powered by the expansion of air came from an outside source, that source was probably Cayley. On the other hand, we have no reason to think that the closed-cycle design which so distinguished Stirling's machine was anything other than his own.

Cayley's life was undoubtedly a full one and, as such things go, appears to have been happy, save for three things. One, obviously, was his inability to perfect his air engine. The second, resulting from the first, was his failure to find a first mover for his aeroplanes. The third appears to have been his wife. When very young, he married Sarah Walker, the strikingly beautiful daughter of his tutor. Sarah was clever as well as beautiful, but, possessed of a violent and malicious temper and, given her social position near the top of the Yorkshire tree, was in a position to make life miserable for a lot of people. In this she appeared to be opposed to her husband, for George Cayley's overriding concern was with the welfare of his fellow men and women. All his scientific endeavours were directed to this end, as were most of his other activities. He was no democrat but, within the bounds of his estates, a benevolent autocrat who would take whatever measures were needful to secure a better life for his people. The grand scheme, which he promoted via an act of parliament, to drain the Derwent levels, was only one example: its success complete and still yielding benefit nearly two centuries later.

Reading of Cayley's doings, there is a temptation to see his extraordinary level of activity as being driven by a need to get away from the wife – and there is some evidence for this. But how to reconcile this with the fact that in his eightieth year he was still writing her love poems? And it would be good to know her side of the story, and to learn why, even as a young and lovely woman in a prestigious position, she insisted on smoking a pipe in public.

CHAPTER 4

John Ericsson's Caloric Engine

The US Navy Board could be forgiven for being wary of John Ericsson. He was big, handsome, loquacious, self-confident and knowledgeable about marine engineering, but he had been associated with a quite spectacular event which, even though the US were not at war, had caused the deaths of some very important members of the Board – and very nearly of the President himself.

Before moving to the US in 1839, Ericsson had done time in a debtors' prison in England – though that wasn't held against him, for in the absence of any legal framework for bankruptcy, any partnership in a business which went bust might have this consequence. Sprung from prison – by whom isn't known – Ericsson took to designing ships' screw propellers, then a novel and rather doubtful idea. Despite the obvious advantages of propellers over the cumbersome paddle wheel, Ericsson was rebuffed when he tried to interest the British Admiralty. (The Sea Lords of the Admiralty had all spent their lives on sailing ships and were doubtful about the future of steam ships in general, let alone those propelled by anything as counter-intuitive as a propeller.) The idea was taken up, though, by a rich American captain and shipowner by the name of Robert F. Stockton. Stockton commissioned Ericsson to design a propeller for his new ship, the *Robert F. Stockton* (he was evidently Ericsson's equal in egotism), which was built by Laird's shipbuilders and later crossed the Atlantic under steam.

Stockton had urged Ericsson to move to the USA, where capitalist enterprise was unhindered by the conservatism which he said characterised the United Kingdom. (Given that Great Britain was by then the greatest industrial power the world had ever known, this may seem surprising – but these things are relative and dependent on your point of view.) This Ericsson did, to find a society in which the diffidence which characterised (some of) the British was unknown and Ericsson's self-promotion was regarded as the justifiable confidence of a superior talent. Given Stockton's advocacy and his connections, Ericsson had access to capital and government, which were even less distinct than they had been in Britain.

Ericsson was charged with the design and building of a new screw-driven warship of 700 tons, to be called the USS *Princeton*. He worked on this for three years until it was launched and commissioned. The ship proved to be a great success, and in 1843 entered a race against Brunel's SS *Great Western*, which it won, being then acclaimed as the fastest steamship afloat. This, naturally, enhanced Ericsson's profile in influential commercial and governmental circles – though Stockton, who had political ambitions, hogged most of the publicity.

The *Princeton* mounted a single, mighty gun of 12-inch calibre and had a novel system to control the recoil, which

Ericsson was widely credited with designing. (Recoil was always a problem with ordnance, and with naval ordnance in particular. Since Newton's Third Law of Motion applies par excellence to big guns, something had to take a recoil whose energy was exactly equal to the force with which the massive shell was propelled.) And because the gun was muzzle-loading, the recoil was not the only problem: forging such a massive piece of iron was at the limit of the available technology. There was always a danger that, given the force of the explosion needed to propel a 12-inch shell, the barrel of the gun might burst when it was fired.

Around that time, relations between Ericsson and Stockton became strained. This is not surprising, since both were self-promoting egotists, and Ericsson was unlikely to be content to remain as Stockton's protégé for long. The gun was test-fired without incident, save the cloud of black smoke which enveloped spectators – which was to be expected, given that the propellant used by such guns was still black powder. Stockton resented the credit which Ericsson was given for inventing both ship and gun; he insisted that another, similar, gun of his own design but the same calibre be forged. The dispute was resolved rather spectacularly: it was arranged that Stockton's gun would be demonstrated to assorted members of the US government and naval high command. The matter was taken up by the press and the demonstration turned into a public spectacle. Stockton invited around five hundred guests, including members of the Navy Board and their wives, the Cabinet and the President. He announced, moreover, that the gun would be loaded with double the charge of powder used on the firing of Ericsson's gun. Ericsson significantly chose not to be close by when the firing took place, but his apprehensions were not shared by various members of the Navy Board who were Stockton's supporters. They paid dearly for their support. When it was fired, the gun burst and killed the Secretary of State, the Secretary of the Navy and six others, besides wounding dozens of spectators and blowing a great hole in the USS *Princeton*. The President, fortunately, had stood some distance away. The episode did good for neither Stockton nor Ericsson, but it did ensure that, from then on, Ericsson was one of the best-known engineers in the United States.

He would go on to become a major celebrity and national hero. In 1861, after the Civil War had broken out, the Confederacy converted the former USS *Merrimack* (which had fallen into their hands because it happened to be lying decommissioned in a Virginian yard when war was declared) to an ironclad battleship. The presence of such a weapon threatened to end the Union's blockade of the East Coast. (The Union's warships were still wooden and mostly sailing ships. Their guns were still muzzle-loading cannon which threw iron cannon balls at a velocity sufficient to damage a wooden ship but unable to penetrate an iron hull.) The blockade was important, for roads were still poor and railroads few, and supplies both civil and military on the East Coast were mainly carried by ship. Whoever controlled the sea had a strategic advantage. The newspapers carried scare stories about armoured vessels penetrating far up the Potomac and shelling the White House.

The US Congress recommended to the navy that they respond by building their own ironclads, so Ericsson was commissioned to design one for them. He was still sore at the navy over the *Princeton* affair, for which he had, he thought, been unfairly saddled with blame. But he allowed himself to be persuaded and devised a ship, to be called the *Monitor*. It is an indication of Ericsson's status at that time that he should have been so courted.

The *Monitor* was built in remarkably short order and ordered to meet the CSS *Virginia*, as the old *Merrimack* had been renamed. This it did, in the best American tradition of conflict resolution by standup gunfight, at Hampton Roads, Virginia, on 9 March 1862. It was a fairly inconclusive battle, with neither ship able to sink the other, but it did prevent a defeat of the Northern fleet by the *Merrimack*, which had otherwise looked likely. The Union navy went on to commission several more *Monitor*-type ships, despite their having been shown to be unseaworthy. (A *Monitor* looked like a flatiron – its popular nickname – with a can on top. The can was a turret which could rotate to give a field of fire to its two cannon. But it was small and manoeuvrable, unlike the *Merrimack*. It was not, however, seaworthy, a matter which soon became evident and caused some embarrassment.) In fairness to Ericsson, seaworthiness had not been a design criterion, whereas several of the other criteria precluded it.

The upshot nevertheless was that Ericsson became reconciled to the navy and the navy to him. The last was not surprising since he had achieved heroic status in the eyes of the Northern press. He would go on to design torpedoes and torpedo boats for the navy, though some of the members of the Board would remain wary of him.

—

John Ericsson is the third in our trio of early proponents of the heat engine. He was part visionary, part showman, part practical engineer. He wasn't the first engineer to be a celebrity, but he was probably the greatest of his time – at least on the western side of the Atlantic. And his prominence wasn't because all his machines worked perfectly – some of them didn't work at all – but such was his charisma that he carried people along on his tide of optimism, panache and bluster.

Born in Sweden in 1803, Ericsson was a physically and morally powerful individual whose success was due as much to his forceful personality as to his undoubted inventiveness in matters mechanical. In his early years, his parents were in comfortable circumstances until his father lost most of their money in speculation and had to take employment as a 'director of blasting' in the construction of the great Göta Canal. It looks as though the propensity to take risks was inherited by his son.

Sweden at the time was a much more corporate state than the United Kingdom and the canal was the work of the Swedish Navy. Ericsson and his brother became naval cadets thanks to the patronage of an influential benefactor who recognised Ericsson's nascent talents and engaged him as a surveyor cadet at the age of fourteen. Having trained as a surveyor, Ericsson left the navy at age seventeen and transferred to the army (the corporate state was evidently flexible), which sent him to northern

Sweden to carry out more surveys. While there, he is said to have designed and built a heat engine, though little is known of it save that it appeared to run well.

After a few years in the army, Ericsson resigned his commission. Seeking resources to develop his engine, he moved to England in 1826, a country then beginning the second stage of its headlong Industrial Revolution. He exhibited his engine but seems to have been unable to make it run properly and consequently found no backers. The reason given for the failure of the engine is usually that, being designed to run on birch wood, it was unable to cope with the coal which was the fuel of choice in England. This is one of a number of unquestioned but implausible myths which surround Ericsson. Anyone who has experience of getting birch to burn is likely to be sceptical: it is a timber which will not burn well unless it is very dry, and even then doesn't produce a great deal of heat. Coal, on the other hand, can be made to burn very hot indeed but, once the fire is lit, its heat output is easy to control. The real reason why the engine didn't work well is unknown, but that's not surprising: the basic design of Ericsson's engines was not a good one and it took skill and experience to get one to run at all.

Given the defects from which Cayley's engine suffered on account of the corrosive nature of the working gases, one might expect a similar cause for Ericsson's difficulty. But if the early engine was like its successors, that would not be the case, since Ericsson's engine was the clean version of open-cycle engine we mentioned in Chapter 3. As with the Cayley engine, a new charge of air is brought in for each stroke, but the heating is external and the flue gases are kept separate from the working air. In consequence, corrosion and sealing problems are fewer: Ericsson's engines worked much better than Cayley's and suffered far fewer breakdowns.

On first coming to London, Ericsson was not wealthy, but appears – then, as later – to have had little difficulty in attracting investors by his combination of flamboyance and engineering know-how. He went into partnership with John Braithwaite, who had inherited an engineering business from his father and was a skilled draughtsman as well as an experienced practical engineer. The partnership designed and built a railway locomotive which was appropriately named *Novelty*. It competed in the 1829 Rainhill Trials on the new Liverpool and Manchester railway, in which it came second. (You may recall that, as early as 1816, a railway had been built from Kilmarnock to Troon and a Stephenson locomotive had been trialled on it.) A Stephenson engine won the trial. By the late 1820s, it had become apparent to many people that railways would replace canals for the transport of heavy goods – and perhaps passengers. A railway had been built from the great manufacturing metropolis of Manchester to the port of Liverpool. It was a great undertaking and testimony to the belief in the future of railways – even though it hadn't been decided whether steam locomotives were capable of hauling a train of wagons on rails or climbing a gradient. (This was one of the big questions to be answered by the trials. Of course, if they had asked Trevithick, he could have told them, for his engine had done both, hauling 10 tons of iron and seventy passengers a quarter-century before. It is sometimes astonishing how little one part of a small country knew of what other parts were doing.) A trial was arranged at a place on the line called Rainhill, and anyone who had a locomotive machine might enter it. Ten were entered, of which five turned up. (One of these consisted of a carthorse walking on a treadmill connected to the axle of the engine.) The favourite was *Novelty*, which was technologically the most advanced of the five and on the first day reached the unheard-of speed of 28 miles per hour. Alas, *Novelty* had boiler problems, was unable to complete the course, and *Rocket*, which the Stephensons had entered, was declared the winner.

The Braithwaite–Ericsson partnership built another two railway locomotives, and then the steam engine and boiler for the ship *Victory* (financed by Felix Booth, a gin distiller) in which John Ross proposed to find the Northwest Passage by a combination of Navy enterprise and high technology. (He was the first in a long line of British explorers who were unable to understand what we would today call *appropriate* technology.) The expedition was a failure and the steam engine was dumped on a beach, but the publicity for the Braithwaite–Ericsson firm was great. They would continue to exploit the new technology of high-pressure steam in various inventions, of which the most successful was a series of fire-engines (i.e. engines for putting out fires). As so often happens, the fire-engines met strong resistance from people with a vested interest – in this case, in buildings burning down. Despite its high profile, the business was unprofitable; in 1835, Ericsson was declared insolvent and cast into the King's Bench debtors' prison.

Ericsson was nothing if not prolific. Besides the above and a host of unrelated inventions, he spent much of his working life in developing and perfecting his Caloric engine. It was he who named it and, as the name implies, he was a firm believer in the Caloric theory of heat. In London, in 1833, he demonstrated a Caloric engine which had a 14-inch piston, a 6-inch stroke and was rated at 5 horsepower. It ran to the satisfaction of most onlookers, and in the general enthusiasm its defects were at first overlooked. Its method of operation was described by Ericsson. There were two cylinders: one, as in the Stirling engine, holding the power piston and the other the displacer, the latter shifting the air from hot to cold end of the cylinder. After the incoming air had done its work, and consequently gained heat, it passed through the regenerator, where it transferred its heat to the incoming air. The working cylinder was pressurised and the engine had a regenerator – Ericsson's first mention of it.

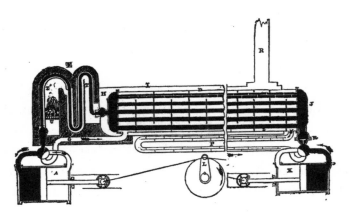

Schematic diagram of the Ericsson engine from 1833.

The engine was well-received. The words of John O. Sargent in the US press were typical of the public excitement it generated: 'The engine will prove the most important mechanical invention ever conceived by the human mind, and one that will confer greater benefits on civilized life than any that has ever preceded it.' A surprising number of eminent persons in England were carried along on the tide of enthusiasm and there was demand for lectures by experts who would explain how it worked. (None of the explanations have survived, which is a pity, for they would have made interesting reading.) But among scientists and engineers the praise was muted, and Brunel for one was not impressed. Claims such as Sargent's were just too extravagant and contrary to the experience of many practical men. The problem was that, lacking an alternative theory of the nature of heat, their opposition lacked force. It would be less so once Joule produced his paper on the mechanical equivalent of heat and showed the relationship between heat and force. Also, practical engineers could see that it worked at very high temperatures – and such temperatures were inimical to some of the things the engine would depend on for long-term continuous operation. The piston seals would break down in the heat, the lubricants would burn or evaporate, and – ultimately – the iron of which the engine was made would begin to oxidize.

Most of the descriptions of how the engine worked were couched in terms of Caloric – and the assumption underlying them was the same as that which Stirling held: that Caloric could deliver work indefinitely without itself suffering any diminution. In a letter to *Mechanics*

Magazine in January 1834, Ericsson explains how his Caloric engine works. The major premise of his argument is that it is not 'necessary to the effect produced that the heat should be absorbed or destroyed, or in any way diminished in energy'.

He gave his opinion that all the heat losses were merely incidental and, not being intrinsic to the operation of the engine, might with care be eliminated. His biographer remarks that his fellow Swede, Professor Llarvefeldt, supported him in this and averred that 'there was nothing in the accepted theory of heat to prove that a common spirit-lamp might not be sufficient to drive an engine of one hundred horsepower'. In some of his later pronouncements, Ericsson makes it plain that he subscribes to the idea of Caloric. He thinks that when the engine is started up, the regenerator becomes charged with Caloric and the Caloric then produces the work of the engine over and over again, its only diminution being through incidental losses. In a letter written to some of his associates, to Messrs Stoughton, Tyler and Bloodgood, in January 1855, he says, 'more motive power may be obtained from a mess of metallic wires of two feet cube (the regenerator) than from a whole mountain of coal as applied in the present-day steam engine'. At another point he alludes to 'the principle which compels metallic threads to yield more force than mountains of coal'.

By far Ericsson's most influential early supporter was the great Michael Faraday. One of the giants, Faraday was of lowly social origin and self-taught in science. His base was the Royal Institution, an organisation second in scientific prestige only to the Royal Society, and charged with presenting and interpreting the discoveries of the day to a popular audience. This it did in various ways, but most famously by its Friday evening lectures, which were open to all and well attended by all social classes. It was announced that one Friday lecture would be on the subject of Ericsson's engine. Given Ericsson's notoriety and Faraday's fame, the hall would be packed. Faraday himself described how, having all the previous day been thinking about Ericsson's engine, he came to the conclusion, shortly before the lecture was to begin, that he couldn't really say why it worked at all. His impulse was to cancel the lecture, but he was dissuaded. Faraday, ironically, confined his talk to explaining the regenerator, which he said was an apparatus which allowed heat to be used over and over again. Evidently even Faraday at the time believed in Caloric, or something of the sort. (If nothing else, this indicates the distance in terms of understanding between his era and that of succeeding generations.)[1]

From the point of view of the audience, the lecture was a huge success and the thing became the talk of London society. It did Faraday no harm, but it was a blow for Ericsson. Being Ericsson, he rolled with the blow and bounced back as usual. It is said that ever after on Friday evenings the secretary of the institution was required to stand on the stair at the entrance to the lecture hall, to catch the speaker lest he try to flee. (This was actually in reference to another speaker, but it makes for a better story if told of the great Faraday.)

---

1   L. Pearce Williams, *Michael Faraday,* p.330.

Soon after Ericsson moved to New York City, he was introduced to Cornelius H. DeLamater, with whom he would form a close friendship and a lifelong association. DeLamater was part-owner of the DeLamater Iron Works, a large and rapidly growing iron-founding, shipbuilding and general engineering company at the foot of West 13th Street on the North River. Ericsson brought most of his many commissions – for iron-hulled, propeller-driven ships, among other things – to DeLamater. DeLamater in return financed Ericsson's many inventions and provided him with facilities to develop his ideas, first to patent and then to production. It was a perfect symbiotic relationship. The works became renowned as a mecca for inventors and promoters of inventions. Ericsson would design his future Caloric engines at the DeLamater works, which would then execute his commissions and market the products. In the ten years to 1850, Ericsson would design and build eight Caloric engines. This, remember, was at a time when horse-drawn stagecoaches still ran in New York City.

In 1851 Ericsson produced a 60-horsepower engine, made at the DeLamater works, which he took to the Great Exhibition of that year in London. This was a much more sophisticated machine than his earlier patent and most of the defects which had plagued the previous engines had been remedied.

The engine employed two pistons, one above the other, each working in its own cylinder but carried on a single piston rod. The lower, larger piston is called the working piston and the upper, the supply piston. Very roughly, when the air below the working piston is heated, it expands and drives the lower piston up. Since the lower piston is connected to the upper, that goes up too – and compresses the air above it, which is forced down and delivers another charge to the working piston. When both pistons are at the top of the stroke, valves open which vent the cylinders to the outside air; other valves allow a fresh charge of air in and the cycle repeats. The exhausting air passes through the regenerator, leaving most of its heat behind to be collected by the incoming air, thus enabling the engine to use 'the same amount of heat over and over again', to quote a magisterial article by Professor William Norton of Yale College.[2] Obviously Professor Norton hadn't read Mr Joule's paper on the mechanical equivalent of heat – or didn't think much of it.

This engine, or a version of it, was to have some astonishing consequences. Ericsson promoted it with his usual panache. Then, in his own words:

> The regularity of action and perfect working of every part of the experimental thirty-inch engine, completed in 1851, and above all its apparent great economy of fuel, induced some enterprising merchants in the latter part of 1851 to accept my proposition to construct a ship for navigating the ocean, propelled by paddle-wheels actuated by the caloric engine.

---

2 Prof. W. Norton, *American Journal of Science*, 1852.

Ericsson was quick off the mark: within nine months, the Caloric Ship *Ericsson* was sitting at the dock, her paddle wheels turning.

The engines (she had four) were prodigious. Ericsson seemed to have no qualms about scaling up his design. (We now know that what works at one scale doesn't necessarily work at a larger or smaller.) The working cylinders of his ship's engines were 14 feet across and the two pistons together weighed upwards of 50 tons. (Ericsson would have had them bigger, but the DeLamater foundry were reluctant to cast something of that size – and the people at DeLamater weren't exactly timid.) Ericsson oversaw every detail of both design and execution. And the engines worked. It remained to be seen whether they would yield the savings in fuel which he predicted for them. If they did, they would revolutionise the world's shipping industry.

Consider the difference: a steam ship must have space for her engines; she must also have space for her boilers, for fresh-water tanks and for enormous amounts of coal. She must carry a large staff of expert and energetic engineers, as well as stokers working like slaves in horrible conditions of heat and dust. The Caloric ship will have little coal and few stokers for, according to Ericsson, the heat of one small fire, since it is repeatedly re-used, will suffice for a long journey. She will scarcely need an engineer at all, for there is so little that can go wrong with her. And she will need water only for washing and drinking, for the engine will need none. All of the above were attractions, but what caught the attention of the shipowners most were the figures which Ericsson gave for coal consumption and cargo space. They could see their profits increasing prodigiously: they could also see themselves being left behind if they didn't get on the bandwagon.

A trial run to Washington was planned, the choice of destination no doubt being prompted by Ericsson's flair for publicity. The members of the Virginia Legislature were invited to a reception on board. Ericsson made a speech. They sent Ericsson an invitation to a return bash – but he had left by the time it arrived. He was a busy man, after all. His return to New York was a triumph.

The triumph wasn't unalloyed, though. There could no longer be any doubt that the engine worked: it just didn't work quite as well as Ericsson said it would, either as regards the speed of the ship or as regards the consumption of coal. Ericsson, of course, pointed out that this was only to be expected: it was, after all, a prototype – and might be expected to come up to expectation after a few alterations. This was a serious matter, though, for his investors had invested purely because they foresaw large profits accruing from exactly those factors.

As late as 1873, Ericsson was replying to critics, saying that the greater-than-expected consumption of coal was due solely to losses of heat due to radiation and the lower speed was irrelevant, since the fuel consumption of the Caloric engine was not related to its speed and the ship could be made to go faster without consuming any more coal. It is apparent that he still believed that the source of the work of driving the ship was the Caloric residing in the regenerators – and Caloric did not diminish with use!

There is a sad end to this story. In March 1874, after alterations to her engines, the ship made a trial trip down New York Bay. A little more than a month later, she made another. All was well, the day was fine, and she was making 11 miles an hour, using very little coal, when she was struck abeam by a sudden gust of wind. Some fool had left open a row of scuttles on her starboard side. She heeled over, water rushed in and she sank within minutes. Ericsson wrote of his anguish, for she had met all his – and, more importantly, his investors' – expectations. No lives were lost and she was subsequently refloated, but the Ericsson luck appeared to have run out and with it his investors' confidence. They withdrew their support and the ship was re-engined with common steam engines. After various mundane employments, she ended her days in the Pacific, carrying coals for the British government.

This was the sort of setback which, for a normal individual, would have been the occasion for despair – and in some cases, for suicide. But Ericsson, as will have become apparent, was a far from normal person. We know that his faith in the virtues of Caloric did not diminish and he is recorded as saying after the sinking that he believed that another $50,000 spent in developing the regenerator would do the job. But failing that, he was determined to 'return to my original caloric engine . . . it cannot fail to be sufficiently useful to save its author from starving'. It certainly did: over the next few years he commissioned and sold thousands of small Caloric engines which were employed for the next half-century in a host of lowly tasks. Those they executed largely free of trouble, for the defects which were so grievous in the large engines were mostly absent in the small. Starvation would not be a problem. We shall return to these in the next chapter, but here we will look at Ericsson's most prescient invention.

—

In a contribution to the Centennial Exhibition, held in Philadelphia in 1876, Ericsson looks at the possibility of using the sun's heat to drive an engine: both a steam engine and an air engine. He credits Sir John Herschel with first having the idea that solar rays might be concentrated, in a series of experiments carried out in Cape Town in 1838. Herschel contrived to direct solar radiation so as to boil water – even to roast meat. Contemplating this, Ericsson observed that if the sun can be made to boil water, it can be made to power a steam engine. He described the elements of a solar machine being a concentrator, a steam generator and an engine – in this case a simple steam engine. Ericsson's invention would later be built, in New York, in 1883. It was an ungainly looking apparatus whose defects are obvious on a cursory inspection. A much more elegant thing is his solar hot-air engine which, we are told, was built in New York in 1872.

A surprisingly modern-looking machine, it consists of an air engine set in a concave concentrator disc, the hot end being the focus of the reflector. Ericsson tells us that the reflector is not parabolic, which would cause it to concentrate the sun's rays too narrowly: it is an irregular curve which will spread the heat across the hot end of the engine. The lower body of the engine is to act as a radiator to take off the heat which, he says, has not been converted into work. (He still doesn't understand that a flow of heat through the engine is essential to its operation.) At the lower end, the power piston is connected to a linkage which transfers the reciprocating motion to a wheel.

The device described is clearly a prototype, built to demonstrate the hypothesis that concentrated solar radiation can be made to drive an air engine. It is not at all evident how the power generated by such a machine might be made useful: it is going to be very awkward to transfer the motion of the wheel so as to drive any mechanism, given that the reflector must be rotated to follow the motion of the sun over the course of a day. But that's the kind of thing which an engineer of Ericsson's ability would no doubt take pleasure in devising.

Ericsson describes clearly how the engine will work: how the heat will flow through it and how the pistons will move. But what is astonishing is that what he is describing is not an Ericsson engine: it is an engine which, in all significant respects, is the engine which Stirling presented in his 1816 patent application. He even explains that the displacer piston (he calls it the exchange piston) will in effect function as a regenerator.

Ericsson has a vision of all the hot places on earth being supplied with abundant power, thanks to Solar Stirling. He doesn't think of it in terms of electricity, which is then in its long infancy, and the linear alternator hasn't been invented. But he sees correctly that freely available solar power will be of inestimable benefit to humankind, and he sees how a Stirling engine might make it possible. He is a little over a century premature.

Ericsson had another, different concern which impelled him to seek to harness solar power. He was concerned about energy security; specifically, he foresaw a time when the world's coal supplies might be exhausted, and he was seeking an alternative. Since our problems are likely to arise more from the overabundance of fossil fuels, he was off-message there – but the distance of his vision is impressive nonetheless.

It isn't easy to trace Ericsson's motivations in his many activities. Certainly, he was arrogant and bombastic, egotistical and assertive. But, judged over a long life, he cannot be said to have been self-seeking – at least in the way that ordinary men seek wealth or fame or power. He was a poor man when he began and he was not a wealthy man when he died. He was careless of his patents and cheerfully assigned or neglected them, leaving lesser men to benefit by them. There is evidence that he had an eye to the general good likely to accrue from his inventions, but this seemed to stem neither from sentiment nor from theory, but to be an incidental consequence of progress itself. He was certainly not a saint, but he was worthy to be up there with Stirling and Cayley.

CHAPTER 5

Near Death

So far, the people who have figured in our account of the rise of the air engine have been exceptionally inventive and unusual people, but none of them have been great theoreticians. They have all been practical engineers and, in as much as they thought about why their engines worked, they all embraced a theory which would later be shown to be false. We might reasonably ask how this could be – and more: we might ask how three individuals with false theories could invent and develop such engines while other people did not, even though the latter persons both had an interest in making such an invention and had ideas about heat and work which were closer to being correct.

Our first question is, I think, the more difficult. It is never easy to say how anyone came to invent anything and explanations tend to be dry-as-dust accounts of antecedents and influences, and may or may not be plausible. My inclination is to look at the people rather than the intellectual milieu. Obviously, they were all people who found engines interesting – but so did millions of other people who didn't invent very special engines. It isn't easy to see precise parallels among Stirling, Cayley and

Ericsson, for they were very different people. What they did have in common was that each of them seemed to have a high level of belief – or self-belief: a conviction that what they had to do was what they *ought* to do. The manner in which this manifested itself was quite different, but there was a lot in common. And there was the pervasive belief in Progress, part of which was the conviction that technical invention would redound to the good of all. (That was obvious for Cayley and Stirling, but it was present in Ericsson too, even if it was overlaid by all the bombast.)

One possible avenue is to attribute it to the zeitgeist: the spirit of the age, which chose the three of them as the vehicles of its self-articulation. I expect most readers will think this just sounds like nonsense; I do myself. But it was to be a common idea throughout the nineteenth century and into the twentieth. It was highfalutin metaphysics and it came from Georg Wilhelm Friedrich Hegel, a miserable German philosopher who had some pretty daft ideas which were nonetheless influential and became part of orthodox thought in many circles (though few folk actually *read* him). Hegel argued (at enormous length) that all reality was one spiritual entity which expressed itself in human history in a sort of logical stagger called dialectic. The theory was called dialectical idealism. Marx turned it on its head, saying the only reality was material, hence dialectical materialism. (If you don't think this was influential in the most basic way, you need to get out more.) The idea of dialectic made a comeback in the 1960s, thanks to a philosopher called Thomas Kuhn, who showed that the history of science proceeds dialectically

– so the daft idea is not too distant from our concerns here.

The history of science is littered with examples of people who seem to have had no connection with each other or with intermediaries, but who come up with the same novel ideas at much the same time. To admit this, you don't have to get metaphysical: it's probably enough to recognise that scientists rarely exist in isolation, that even hundreds of years ago they published what they were thinking about in journals and books, and there therefore arose a community of ideas which different people could tap into. That is probably a sufficient answer to our first question. The second one is harder.

By the middle of the nineteenth century, many scientists were interested in air engines. Among the cognoscenti the Caloric theory had been pretty-well discredited. The flamboyant Count Rumford had started digging Caloric's grave with his observations of cannon-boring; Sir Humphry Davy contributed a few shovels' worth with one of the simplest and most effective experimental demonstrations in the history of science: Lord Kelvin (William Thomson) described how 'Sir Humphry Davy, by his experiment of melting two pieces of ice by rubbing them together, established the following proposition: "The phenomena ----- are not dependent on a peculiar elastic fluid for their existence . . . Caloric does not exist." And he concludes that heat consists of a motion excited among the particles of bodies.'[1] In the same paper, Kelvin refers to the discovery by J. P. Joule in Manchester in 1843 that it was

---

1 William Thomson, 'On the Dynamical Theory of Heat', *Transactions of Royal Society of Engineers*, March 1851.

possible to determine an equivalence between mechanical motion and heat.

As we have seen, Cayley and Ericsson stuck with the Caloric theory in explaining their engines; the theoreticians soon abandoned it in their many discussions of the heat engine. But even without Caloric, the heat engine would be a hot topic for years to come. Joule and Kelvin would write a paper on it which they delivered to the Royal Society; Sir William Siemens wrote one for the Institution of Civil Engineers; I. K. Brunel would write one, so would Sir Goldsworthy Gurney.

W. J. M. Rankine would deliver a paper to the British Association in 1854, in which he made suggestions for improving the engine. He would collaborate with the Glasgow engineer and shipbuilder J. R. Napier and produce a design for an air engine which was by far the most sophisticated yet. Theoretically, it should have far surpassed any existing engine, air or steam, but it soon disappeared from view.

A glance at the drawing will show the reason: the practical difficulties in the way of realising such an engine and of getting it to work would have been insuperable. (If in any doubt about this, consider how the long, finger-like displacer rods in their narrow slots would behave through the wide differentials of temperature necessary to make the thing work.)

The flurry of theoretical studies had little practical effect, not least because many of the commentators, such as Kelvin, had other more important fish to fry. (In his case, transatlantic telegraph cables.) So little effect, indeed, that as late as the 1870s F. H. Wenham, who like

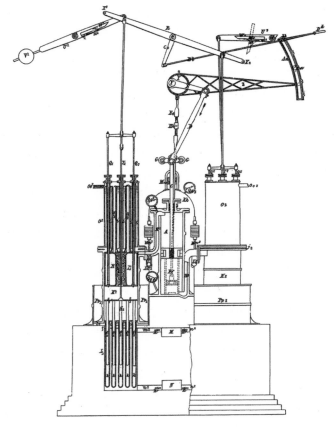

Engine proposed by Rankine & Napier.

Cayley needed a machine to propel his aircraft, would propose a new engine based on Cayley's dirty open-cycle model (see Chapter 2) and, though not espousing Caloric as its operating principle, included a regenerator.[2] In his paper, the author mentions that in the preceding half-century, 'upwards of 250 plans have been brought out

---

2   Conrad Cooke to the Institution of Mechanical Engineers, 1873.

for the application of air expanded by heat as a motive power'. It would not, however, be the theoreticians who would drive things forward but the practical engineers, and the advance of the air engine in the later nineteenth century would embody little in the way of new thinking. If anything, it would go into reverse.

For the sake of clarity, from now on we will use the expression 'Stirling engine' for all the engines which work on the principle of alternate expansion and contraction of a fluid. That doesn't imply any inaccuracy, for, after Ericsson, almost all such machines conformed to the pattern of either Stirling's twin-cylinder model engines or the single-cylinder engine of his 1816 patent (for a diagram, see p. 42). To gain a perspective on what happened next to the Stirling engine, we shall pause to consider what life was like in the later nineteenth century, with regard to what made things work: specifically with regard to electric motors and internal combustion engines.

The electric motor and the internal combustion (IC) engine are the ubiquitous and invisible workhorses of our society, and without them many of the things we take completely for granted would not occur. Our houses would not be built or heated, our clothes would not be made or washed or dried; we would have no running water in our buildings and our sewage would not be removed. Our cars would not run, nor would our buses or our trains. What entertainment we have, we would mostly have to make for ourselves. Our food production would depend on the muscles of people and animals, so would its distribution. The list is long.

This remarkable fact is obvious in the case of the

internal combustion engine; less so as regards the electric motor because electric motors are, by and large, invisible. They, together with IC engines, are, to use the terminology of two hundred years ago, our *first movers*. They make things go. The machines which build our houses have electric motors built into them: the diggers and the bulldozers, the forklift trucks and the concrete dumpers. The drills and the saws all have electric motors, either to drive them or to control them. Our clothes are made of yarn which is spun by electric motors and the cloth is woven by them; the clothes are sewn by sewing machines which are driven by them. Record players, CD players, computers, television sets, cinemas: most of the technology of entertainment depends on electric motors. Most people can go through their lives without ever being conscious of seeing an electric motor, but they are there all the same.

Thinking about this, it is hard to imagine a technologically advanced civilisation which doesn't have electric motors, yet society in Europe, America and some of Asia in the later nineteenth century – in its urban manifestation at least – undoubtedly qualified as a technologically advanced civilisation and it had no electric motors. It had steam engines and some waterwheels and windmills – but none of these are exactly suitable for driving your sewing machine or your dentist's drill. There was accordingly an open field which might be colonised by small-to-medium-sized alternative engines. Ericsson saw this; so did a number of other people, in the US and in Europe.

Of all the necessities of civilised life, clean water is undoubtedly the most necessitous. In many parts of the world the only accessible source of water is the artesian well. The problem with a well is that the water must be lifted out of it, for which a pump of some sort is required. And water, once lifted out of the ground, must be distributed. If the building to which it is to be distributed is more than a single storey in height, the water must be pumped again. The alternative, of course, is to bring the water by pipes or aqueducts from a higher source, but with a few exceptions the rapidly growing towns and cities of Europe and North America in the later nineteenth century had neither the time nor the resources to imitate the Romans, and so they fell back on pumps. Since a pump needs something to drive it, that created an enormous market for a first mover, i.e. a machine to drive the pump. Windmills and donkeys had long been favoured as first movers for such purposes, but neither was exactly suited to urban application. Neither was the steam engine, for it demanded constant, skilled attention and nobody wanted a potentially explosive boiler in the cellar.

Enter the Ericsson pumper. Following the failure of the Caloric Ship *Ericsson*, its inventor seems to have concluded that the future for him lay in smaller engines. The motive behind this may have been financial, for many of the Ericsson backers lost a great deal of money over the ship. It may also have stemmed from Ericsson's appreciation of the problems arising from scaling up the engine: defects which were negligible in smaller engines could become significant in the larger ones. Getting a 14-foot piston to make a gastight seal to the wall of a 14-foot cylinder using leather washers springs to mind.

Smaller engines suitable for pumping and other duties were built to Ericsson's design and widely marketed. By 1861 he is found writing to Abraham Lincoln (no less) about his pumper engine, saying that at the time of writing more than a thousand had been sold. The engine was an open-cycle machine and was made in a variety of sizes: cylinder bores ran from 6 to 24 inches. Some twin-cylinder machines were built with 40-inch cylinders and even a mighty 48-inch, which was sent to power a Cuban sugar plantation mill. One imagines that some of the scale defects would have been apparent in the latter, but nothing is now known of it. (The owners of Cuban sugar plantations appear to have been somewhat neglectful of their duties to historians.)

The engines were simple and reliable: the advertising often refers to both attributes. Since the only skills needed to run the things were those possessed by anyone who could light and tend a small fire and squirt an oil can, both must have been true. The machines were undoubtedly inefficient, in that they used a lot of fuel for very little effect; but fuel in the form of either wood or coal was plentiful and cheap. And they needed to be reliable, for a job like pumping water was one which must be done constantly and without interruption for service or repair. The engines were, though, very heavy: made of massive iron castings and forgings, some of the earlier models were estimated to weigh as much as 3 tons per brake horsepower. (My lawnmower's little petrol engine, which I can pick up with one hand, delivers about three brake horsepower.)

Ericsson seems in time to have recognised the defects of his open-cycle design, for in his 1880 patent for a

pumping engine he adopts a closed-cycle design not significantly different from that of Stirling's 1816 patent.
It has been suggested that he may have been influenced
in this by the success of the Walden engine (see below),
which was a much lighter and altogether more advanced
design.[3] However that may be, Ericsson's new engine
met a very favourable reception and it sold in large numbers until his death in 1889. The engine is essentially an
upside-down Stirling, with the heater underneath and a
simplified link arrangement of levers connecting piston
and displacer with crank and flywheel. The regenerator
seems to be fading by this time, for little is made of it. As
we shall see, later engines by other designers would omit
the regenerator altogether: the engine worked without
one and, if expectations of thermal efficiency were low,
its absence would not be noticed. Besides, by the end of
the nineteenth century, belief in Caloric had pretty-well
faded – and without it, there was not even the semblance
of an explanation of how the regenerator worked.

The protection offered by a patent was far from complete, either in the USA or in Europe. A minor variation
in design could be sufficient to act as a defence against
a suit for infringement of patent. In consequence, many
new versions of the air or Stirling engine appeared in the
later nineteenth century on both sides of the Atlantic.
Very few of these were any improvement on Stirling's
original designs and many were less efficient than them.

---

3   Brent Rowell, *The Rider-Ericsson Engine Company*, p.18.

Here is Brent Rowell, of the Coolspring Power Museum, in Kentucky, USA: 'Most hot air engines of the day, including Ericsson's, differed little from the early beta and gamma configurations of Robert and James Stirling, and these differences more often than not resulted in decreases rather than increases in efficiency compared with the original Stirling engines.'[4] There was one exception, though: Alexander K. Rider.

Born in Belfast, Northern Ireland, Rider was the son of an engineer, Job Rider, who shortly before Alexander was born patented a rotary steam engine which was claimed to be more economical than the traditional reciprocating engine. The design was licensed to Claud Girdwood of Glasgow who, in the early 1820s, built one and installed it in a small ship called the *Highland Lad*. The engine seems to have failed to meet expectations and Rider was called in. What happened then is unclear: the interest of this lies in the fact that, at the time, James Stirling was an apprentice at Girdwood & Co., so Rider senior may have met him and learned of his brother's interest in air engines. James later fitted an air engine of his own and his brother's design to the *Highland Lad*. Girdwood's must be commended for their persistence in seeking novel first movers – but their previous experience may go some way to explain their decision, when the Stirling engine failed to deliver, to replace it with a more traditional and reliable steam engine.

Alexander Rider emigrated to the USA in 1841 to join his elder brother William, who was by that time an

---

4   Brent Rowell, *The Rider-Ericsson Engine Company*, p.19.

Plate 1. Dr Robert Stirling in old age: much-loved pastor and engineer.
(History and Art Collection / Alamy Stock Photo)

Plate 2. *Right.* A mysterious machine bought from a stall on the Portobello Road one Friday morning. (Phillip Hills)

Plate 3. *Middle right.* Innerpeffray Library, the oldest public library in Scotland, lending works in science as well as demonology and witchcraft. Robert Stirling's signature can be seen in the Borrowers' Register for 1806. (Duncan McEwan, Innerpeffray Library)

Plates 4 and 5. The Edinburgh Engine (*below left*) and the Glasgow Engine (*below right*). Both were devised and built by young Stirling, presumably with help from the staff of the universities of the two cities, to which he appears at some point to have given the models. (*below left* © National Museums Scotland, *below right* © The Hunterian, University of Glasgow)

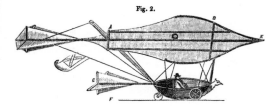

Mechanics' Magazine,

MUSEUM, REGISTER, JOURNAL, AND GAZETTE.

No. 1520.]   SATURDAY, SEPTEMBER 25, 1852.   [Price 3d., Stamped 4d.

Edited by J. C. Robertson, 166, Fleet-street.

SIR GEORGE CAYLEY'S GOVERNABLE PARACHUTES.

Fig. 2.

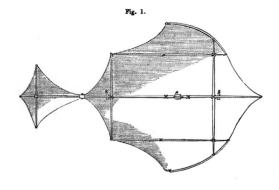

Fig. 1.

Plate 6. *Above left.* Sir George Cayley: scientific gentleman. (Photo Researchers / Alamy Stock Photo)

Plate 7. *Above right.* John Ericsson: engineer and hustler. (Nicku / Shutterstock)

Plate 8. *Left.* Almost an aeroplane: Cayley's glider. (FLHC26 / Alamy Stock Photo)

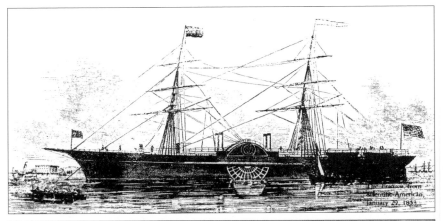

Plate 9. *Top*. The Caloric ship *Ericsson*.

Plate 10. *Above*. Technology at war: the battle between the *Monitor* and the *Merrimack*, which made Ericsson a legend. (Everett Collection / Shutterstock)

Plate 11. *Right*. The Ericsson Sun Motor: a full century in advance of its time.

Plate 12. A Rider–DeLamater engine.

Plate 13. A Jost fan, made in Germany and sold in British India. (Phillip Hills)

Plate 14. An erratic: Malone's large engine.

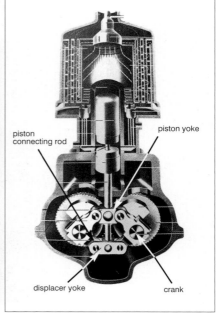

piston connecting rod

piston yoke

displacer yoke

crank

Plate 15. The beginning of the renaissance: cross-section of the Philips Stirling engine.

Plate 16. A brilliant engineer and a very determined guy: William T. Beale.

Plate 17. The free-piston engine as supplied to NASA. (NASA)

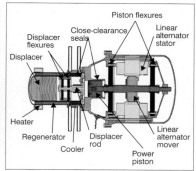

Plate 18. The extreme simplicity of the shell conceals an interior whose realisation embodies vast complexity.

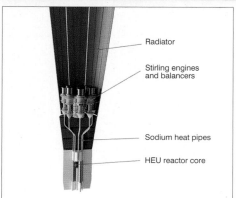

Plate 19. *Above*. KRUSTY in the NASA workshop. (NASA)

Plate 20. *Left*. KRUSTY schematic.

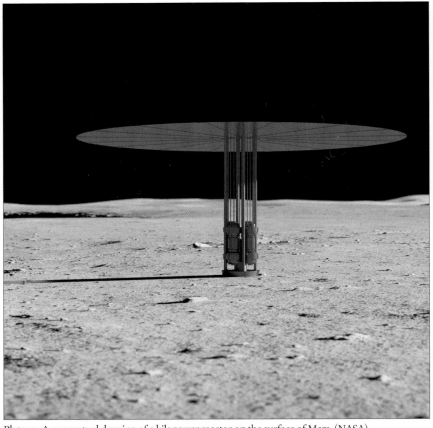

Plate 21. A conceptual drawing of a kilopower reactor on the surface of Mars. (NASA)

Plate 22. A NASA imagining of a Mars base supplied with power by scattered KRUSTIES. (NASA)

established engineer. The brothers continued their father's researches into rotary steam engines, jointly filing an application for one in 1843. Rider filed another in 1864, so the thing evidently had legs. By 1850, we find Rider as foundry foreman at the DeLamater Iron Works. Before we proceed with Rider's history, it may be germane to say a few words about mechanical engineering in the USA in the mid-century, and the professional and commercial relationships among the people engaged in it.

The first thing to remark, is the material of which the machines were made: cast or forged iron. (The two are different, but, for our present purpose, not by much.) This is stuff which most people nowadays wouldn't recognise if they held it in their hands. It is very heavy and very hard and if you want something made of it now, your best course is to look to China or India. But by the mid-nineteenth century it was ubiquitous in Europe and North America. Railway engines and railway lines were made of it; so were bridges and sewing machines and boot scrapers and hammer heads. (It would be largely replaced by the new wonder material, steel – another form of iron but much stronger, made by a process which relied heavily on the heat exchanger which Siemens patented but which had been invented by our own Robert Stirling in 1816.) Cast iron is made by heating iron until it melts and then casting it in a sand mould. When cold, it retains the shape of the mould and quite a lot of the sand. It must then be fettled to make it smooth and, where necessary, machined on a lathe or milling machine. The process of casting iron is called iron founding and the places where it happens, iron foundries.

There were many iron foundries in and around New York in the mid-nineteenth century. Since almost any machine would involve iron founding, close relationships were formed among the inventors, draughtsmen, builders, founders, machinists and financiers of the growing US industrial base. DeLamater and Ericsson, of course, were prominent among these, as was Rider. We won't go into the relationships, commercial and otherwise, among these men, but the second remarkable thing is how fluid and generally co-operative the relationships appear to have been. Ericsson would receive free accommodation from DeLamater for much of his life, at a value calculated at some hundreds of thousands of dollars, without any explicit arrangement for payment. Rider would work at DeLamater, then move away to set up his own operation, register some patents in his own name, and later build the engines under those patents at DeLamater. In many ways, it was capitalism working at its best and most creative. In 1871 Rider patented a variation on an Ericsson-type open-cycle engine, which was of no great interest, but also that year, in Philadelphia, he patented a new closed-cycle engine which was of very great interest indeed.

A cutaway drawing of the engine is shown, but its modus operandi is not at first obvious. It is a two-cylinder machine, in which each of the long, loaded pistons is a gastight fit in its cylinder. Or at least the top part of the piston fits: the bottom does not, so each piston pushes the air under it up or down. The two pistons are connected by a tube, in which sit a lot of iron plates which act as regenerator. Since the pistons are set at an angle of about

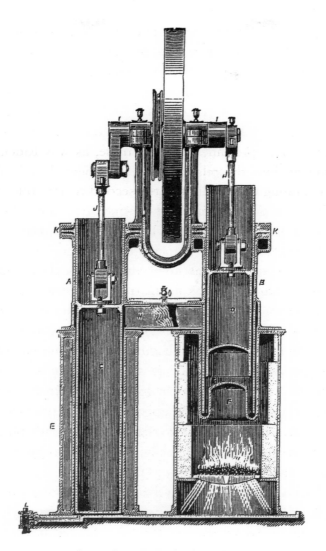

The Rider patent hot air engine.

90 degrees on the shaft, the air in the cylinders is squirted first one way, then the other as the flywheel turns, and at each turn it goes through the regenerator. The bottom of the right-hand piston is heated by a coal fire; that of the left is cooled by a water jacket. When the air under the right-hand cylinder is heated, it expands, forces the piston up and the hot air through the regenerator to the other cylinder (whose piston at that point is rising). When the air hits the cold cylinder, it cools and contracts, and the reduction in pressure cáuses the piston to retract, moving the shaft and returning the air to the hot cylinder to be reheated.

The advantages of this engine will, I hope, be obvious. (If it's not, read the above again and think about it: it will dawn, eventually. But don't worry if this takes some time: it took me about fifteen years.) Firstly, there are only five moving parts, so no complicated linkages to go wrong and consume, through friction, some of the power of the engine. Secondly, both the piston rods are open to the air, so few sealing problems, especially since the upper part of the hot cylinder was cooled by a water jacket to protect the leather piston seal. Thirdly, because of the regenerator, the thermal efficiency was much greater than usual.

The last consideration is curious because in fact the regenerator was put in the wrong place, being at the cold end of the expansion plunger and the warm end of the compression plunger – the two places in which it might be least effective. Nobody seems to have noticed this until Theodor Finkelstein and Allan Organ (see the Bibliography) pointed it out in 2001. The engine was nevertheless very popular, probably because of its extreme simplicity and reliability.[5] It was usually supplied with a water pump attached, but lent itself readily to other sorts of application. And it was long-lasting: there are testimonials which describe Rider engines which have been running for twenty years without breakdown despite minimal maintenance.

—

Variations on the Stirling engine were quite astonishing: literally dozens of different versions were made for as many different purposes. The Rider-engined pumps were used on railways to supply well water for steam locomotive engines: steam locos were very thirsty beasts indeed and they must be supplied with reasonably clean water. Farms used the engines to provide water for animals and for irrigation; towns used them to pump domestic water from distant wells and to supply tanks in tall buildings, as well as to circulate central heating. The engines were widely used in industry: especially in the thousands of workshops too small to justify the installation of a steam engine. There they drove saws, lathes, drills and milling machines, as well as cranes, planers and machine hammers. The engine could be supplied small enough to drive an individual machine or large enough to drive an overhead shaft from which drive could be taken by countershafts to particular applications, as was common practice in larger, steam-driven workshops.

The later nineteenth century suffered from what we would today call an information revolution. Literacy was

---

5 Finkelstein and Organ, *Air Engines*, p.51.

widespread for the first time, there was huge demand for printed paper and consequently for printing presses. The Stirling-type engine was perfectly suited to the latter: it could be provided in just the right size, the office boy could light its fire first thing in the morning and need tend it only a few times during the working day. The engine was in demand for churches: where previously the organ had depended for its air on a boy doing the pumping (from the 1890s he need only get up in time to light the fire). It could supply secular music too, for the air engine was used to drive player pianos and even gramophones. (There are advertisements offering all these things and more, with graphics.)

To continue the list would be tedious, but there are a few applications of the engine which really have to be mentioned, if only because they seem now to be bizarre. It was used to drive dentists' drills – presumably as an alternative to the usual foot-pedal; there is an advert for a room fountain for a large house or hotel, in which a water jet emerges from a bunch of flowers, having been pumped by a Stirling engine beneath; there are rotating shop window displays; there are model boats and tram-cars. The engine was even used to provide lighting: by driving a dynamo for electricity and by giving gaslight by blowing air over petroleum to produce an inflammable gas. (The latter must have smelt disgusting and been a safety nightmare: the Stirling engine which pumped the air was fuelled by an open gas flame from the same source. Starting it must have been interesting.)

Possibly the most improbable use for the engine may have been to drive a fan. The idea of using a heat engine running on a burner to cool the air seems odd, but they were very popular and several different firms offered them. It may help to give an idea of the small-scale use of the engine to look closely at one of these.

The thing, as you can see from the photograph (plate 13), is rather handsome: the body is sheet-iron fretwork and within it the cylinder and its adjuncts are cast iron. The rest of the fan is polished brass. The machinery is driven by a brass burner which will run on alcohol or any inflammable, volatile liquid. (The burner is a most ingenious device in which the fuel is delivered to the jet by the capillary action of an internal wick, but then (I think) propelled through the jet by pressure created by the heat of the burner expanding the air inside the reservoir. There is a little manual pump internal to the reservoir which allows fuel to be pumped into a gallery under the jet for starting.)

The engine is an inverted version of Stirling's 1816 patent: instead of the fire heating the top, it heats the bottom of the cylinder, and the heat is lost through radiation and convection by means of (relatively) massive fins just below the fan, whose air cools them, a neat and ingenious arrangement. (To use a fan, whose purpose is to cool the air, to carry away the heat of its own driving, seems counter-intuitive, but it works satisfactorily, presumably because the cooling effect of the fan is great in proportion to the amount of waste heat.)

The displacer is, of course, in the lower part of the cylinder and the power piston in the upper, with the displacer rod passing through the power piston. The latter is actuated by a small yoke and both yoke and piston rod are connected by a linkage in such a way that the two move in the correct angular relationship. The displacer rod passing through the piston is an obvious source of inefficiency but, provided a light oil film is maintained, seems to cause no problems. The displacer, which is hollow and made of copper, presumably does service as a regenerator.

The fan runs beautifully. It will run on the alcohol which I buy for French polishing, but it will also run on overproof whisky, as I have demonstrated on several occasions. (I have an association with a company which, on account of some long-ago delinquencies, keeps me supplied with such stuff.) Responses to these demonstrations have been mixed, in Scotland, at least: everyone is delighted with the fan, but using whisky to run it seems to some a bit like sacrilege.

The fan has a respectable lineage, which I have been fortunate enough to trace. It was made toward the end of the nineteenth century in Germany by a firm called Jost. Germany wasn't a huge market for fans, but the British in India *were*. A pictorial advertisement from the *Times of India* in 1914 shows, on the left, a British gentleman in full evening dress fretting in the heat at a desk; then, on the right, the same person relaxing in front of a Jost fan. The legends are, respectively, 'No Punkah – no work' and 'Punkah – work'. The punkah is, of course, the fan which would, in the absence of a Jost fan, have been worked by some poor devil outside the room who pulled a rope. The fact of our British imperialist doing his daily work in evening dress does not seem to be regarded as in any way remarkable.

One assumes that the fan, having been sent to India, was purchased by a member of the Raj who did his duty

to Queen Empress in evening dress. Not all the servants
of the Raj were the sort of selfless functionaries described
by Rudyard Kipling, though some of them undoubtedly
were and have been written out of history, which is a
shame. But it is undeniable that for well over a hundred
years there were many British who made simply gigantic
amounts of money out of the Indians. Some of the dutiful
public servants, having spent their lives on a government
salary, chose to end their days in India, which they had
come to love. But the plunderers almost always took their
plunder back to Britain, where they would buy a country
house, in which they could flaunt their wealth in the face
of the less fortunate. They were known locally as Nabobs
and everybody hated them.

Such loathsome persons would typically have their
entire possessions crated up and sent by sea to Scotland
or England. (The Scots were enthusiastic participators in
the plundering aspect of imperialism.) The possessions
would be used to furnish house and estate, and served the
dual function of displaying both the wealth and its exotic
origins. Our fan can't have ranked highly in the matter of
display, but it was crated up along with the rest and only
unpacked when it came to rest in the county of Angus
in north-east Scotland. Since the fan, though handsome,
was unlikely to impress the douce folk of Angus, its only
value would have been functional. But, since keeping cool
is not generally a problem in Angus, it was relegated to
a barn or outhouse, where it lay for a century or more in
company with pigeons.

Time passed and fortunes changed and the house
and estate were put on the market. The more valuable

artefacts were sent to London, to Christie's or Sotheby's, and the rest went to Taylor's, the local auctioneers in Montrose, where one Friday evening I spied something interesting among the junk. It was very old and very dirty and covered in more than a century's worth of bird droppings. I bought it for a few pounds (I was the only person in the auction who knew what it was) and took it home. I cleaned the droppings and dismantled it. The cylinder turned out to have been scored, but I borrowed a cylinder hone from our local garage, set the cylinder up on the lathe and honed it. I reassembled it, using as reference the photos of the linkage I had taken. The burner, in fact, proved more troublesome than the engine, for it is made of brass with a finely fretted gallery, and where the gallery screws into the reservoir it was corroded. But a few months in WD-40 fixed that. It is a joy to see working: the linkage makes a slight clacking noise as it goes up and down, and the fan spins merrily. Run on pure alcohol, it makes no smell; run on overproof whisky, it makes a bit, but nothing that the sort of folk likely to see it working are going to mind.

—

Almost all the machines I have described above would come to be worked by electric motors, save for those which, because electricity would not be available or because higher power would be needed, would be driven by an internal combustion engine. After Rider, though Stirling engines would be built in many varied forms, there was no real innovation for a long time, and the Stirling engine died a slow death, mainly because of the cheapness and convenience of the alternative prime movers. Electric motors would improve in basic design in the first few decades of the twentieth century and lend themselves to a great variety of functions, but their development was slow compared to that of the internal combustion engine. While the Stirling engine was static, the internal combustion engine, both petrol and diesel, was on a perfectly amazing trajectory. In researching this book, I came across a contrast which will nicely illustrate the point.

Toward the end of the nineteenth century, Robinson engines appeared on the UK market. They were fairly primitive, lacking either regeneration or raised pressure and were less efficient than Stirling's original design. Made in small sizes, they were marketed as 'one-manpower' or 'two-manpower' motors. They could be driven by an oil lamp or a gas burner, and they sold in their thousands. By 1920 they could still be had from Messrs Norris, Henty and Gardners Ltd, of London and Manchester.

No doubt Gardners were grateful for the profits and cash flow from the sale of Robinson engines, but they had their sights set on better things, for they could see that in the diesel engine they had the stuff of the future. Gardners made very fine diesel engines and for a long time were at the forefront of engine design. Within sixteen years of 1920, they announced the introduction of their LK series diesels. The LK was a small, high-speed engine whose maximum speed was 2,400 rpm – then an unheard-of rate for a diesel. It had an alloy cylinder block, which meant that it was light for its size compared with other engines. It had a very high compression ratio and an in-line injection pump whose timing could be altered to suit speed and load. It came in two-cylinder units which could be combined to provide an engine of the power required. Its swept volume was a litre a pot, and it was a thing of beauty.

The engine would be put to many uses. It powered the midget submarines which crippled the German battleship *Tirpitz* in 1943, it powered armoured cars in the Western Desert, and the Commissioners of the Northern Lights installed it in the lighthouse on the Isle of May and elsewhere. Gardners' main agents in Scotland, Bowen's, at one time had a sixteen-in-line running a standby generator. But Gardners, being a forward-thinking engineering firm, had their eyes on one market in particular, which no one but they had considered and for which they had the machine. They thought that the day would come when we would have diesel-engined motor cars. And the LK series engines would demonstrate that that was a realistic possibility. The chassis in which the engines were to run would have to be robust, so the cars would have to be large – for even an LK would take a bit of fitting into an existing car.

In 1937 the Lagonda Motor Company had two cars ready for the Motor Show. They had enticed the great W. O. Bentley away from Rolls-Royce, a company which he didn't much like working for after it took over his bankrupt company and produced Rolls-Bentleys. Lagonda got Bentley to fettle their existing M45 model, a fast, racy sports tourer which was a bit rough at the edges and had an old-fashioned Meadows petrol engine. Bentley re-engineered the thing from chassis up. No doubt it came as

126

something of a surprise when Gardners bought both cars off the stand and asked if they could be delivered up to Manchester minus the engines. They installed a four-cylinder LK in one and a six in the other. The six took a bit of hacking, but the four fitted rather nicely. Gardners ran the cars until 1951, by which time they concluded that the market for diesel cars would never develop. They removed the engines and sold the cars, minus engines, to a friend of mine. He persuaded them to let him have the four-cylinder engine, which he replaced in its original car. By a series of accidents, it came into my hands in 1974; I ran it every day for twenty-five years, at which point it had covered more than half a million miles, and eventually sold it. It is still running today, taking a family on holiday each year to France, in the hands of a careful and appreciative owner, more than eighty years after it was built.

I suppose I ought to ask the reader's indulgence for telling this story, but I think it is justified as illustrating the advances which had been made in the design of the internal combustion engine at a time when the Stirling engine had not only not advanced, it had retrogressed. There was, however, one slightly bizarre exception to the general retrogression, and for that we will look to John Malone in the next chapter.

CHAPTER 6

Street-fighting Man

In all the engines we have looked at so far, the working fluid – i.e. the gas which fills the engine and which is expanded and contracted – has been air. If you can remember school physics classes, you might recall that, unlike gases, liquids do not expand or contract when heated or cooled – or if they do, it is by a negligible amount. And unlike gases, liquids cannot be compressed, which is what allows us to have hydraulic machinery. (When you put your foot on your car's brake pedal, the force is transmitted through pipes by hydraulic oil which, being incompressible, actuates the brake callipers.) It is hard to see how, given this, anyone would imagine that a liquid might replace the air in a Stirling engine. But that is what a remarkable man called John Malone did, in north-east England in the 1920s. A *very* remarkable man, Malone was not the sort of chap you might have imagined as an engineer, a scientist, or an inventor of machinery – even judged by the eccentric standards of Stirling, Cayley and Ericsson. Nor, it must be said, does Malone fit the pattern which we have seen, of an inventor whose purpose is first to benefit humanity and only secondly to profit himself. (This may be stretching it a bit in Ericsson's case, but we can let that pass.)

John Fox Jennens Malone was born in Wallsend-upon-Tyne near Newcastle in 1880. In the early 1890s, his father having been made works manager at John Penn and Sons, the family moved to London and John Malone was apprenticed at the Penn works. For a budding engineer, he could not have been apprenticed in a better place. John Penn's was a supplier of high-pressure, high-speed marine steam engines to the Royal Navy. The company was an early exponent of engineering mass-production, through the application of the most up-to-date standards of precision measurement and machining. Penn's was an enlightened and generous employer which offered its apprentices the best possible education in all matters pertaining to marine engine design and production, both practical and theoretical. John Malone would draw heavily on his training at Penn's in later life.

It is rumoured that Malone was in some trouble with the police when he was seventeen or eighteen years old, a rumour which may well have been founded on fact, given that while still officially an apprentice he had a flourishing small bookmaker's business. This did not meet with parental approval, however, and that having been stopped, he left Penn's and went to sea on a tramp steamer.

It appears that Malone was of an aggressive disposition: he was said to have a spear wound from an Arab war in the Persian Gulf and to have collected a total of fifteen knife wounds in brawls, mainly in South American ports. (This establishes the aggression beyond reasonable doubt: you can get into a brawl by accident, but you get into lots of knife fights only if you go out looking for them.) During the Argentine revolution of 1902, while on his way back to his ship, which lay alongside a quay in Buenos Aires, Malone was assailed by five revolutionary soldiers armed with knives. Having been on watch that night, he was armed with a revolver: he shot three of the soldiers dead and wounded two others, but received five knife wounds in return: serious but not mortal. He returned to his ship, leaving a trail of blood. When he reported what had happened, the crew realised that there would be serious trouble, since the wounded soldiers would report the affray. The crew, evidently accustomed to dealing intelligently with emergencies, shifted the ship's gangway away from the end of the trail of blood – which presumably suggested to anyone following that the malefactor had ended up in the harbour. The crew patched Malone up, put him in a boiler suit and, when a search party arrived and demanded to have the whole crew mustered, presented Malone in his boiler suit, reeking of whisky and supported by his mates, as a typical drunken – but innocent – sailor. This catalogue of delinquency may seem irrelevant to the history of an engine: it is not, as we shall see.

Malone left the sea about 1912 and that same year formed the Sentinel Instrument Company, the first of four similarly named instrument companies he would own. We do not know what sort of instruments he made, but it seems likely that they were either navigational instruments or (more likely) engine-room instruments such as gauges and indicators. With a war looming and ships being built, there was a greater than usual demand for such things. The business (and its successors, the Fox Instrument Company, the White Fox Instrument Company and, in

1932, the Malone Instruments Company) were probably profitable: they enabled Malone to accumulate capital sufficient to fund his engine experiments. We are told – by the *Newcastle North Mail & Chronicle* in 1925 – that 'during the past seven years Mr Malone has devoted his life almost wholly to this work, and for two and a half years has had a faithful staff around him in the erection of the new engine'.

Apart from the faithful staff, of whom we know nothing, Malone seems to have been a loner. Since he had no formal engineering qualifications, he would have been barred from membership of one of the professional engineering bodies, which would cut him off from peer-group support as well as the informal communication in which new ideas are examined. He seems to have retreated into relative isolation and became dismissive of forms of knowledge other than those acquired by practical experience. In a letter written in 1939, he says that 'to my amazement I found that my enemies were alleged centres of learning. Universities and the like.' Later he says, 'A study of liquids as mediums in thermodynamics will teach an engineer more about the art of thermodynamics than all the universities on earth, or the memory men who infest them, and knowledge for knowledge's sake is better than their parasitical life.' This rather suggests that Malone had had his ideas dismissed by people who should have known better, and was bitter about it.

—

As a trained and practical engineer, Malone would certainly have been familiar with air engines of the sort described in the preceding chapter. He seems to have conceived the idea – of a Stirling-type engine using a liquid medium – some time during the 1914–18 war. Where this came from, we do not know. Author Robert Sier writes, 'Malone claimed he discovered his new power medium while engaged on research into the possibilities of power without heat . . . when out of his failure he made his discovery.'[1] While that may be true, it doesn't help us much.

The idea of a liquid-filled heat engine, despite being wildly counter-intuitive, was not completely new. It is based not on the behaviour of liquids as most of us know them, but on the possibility of using supercritical fluids. These are liquids which, if temperature and pressure are high enough, exist in a state which is neither liquid nor gas, but partakes of both. Two of these are of particular interest to the designer of a Stirling engine: the supercritical fluid is incompressible like a liquid, which gets rid of the problem of the unwanted internal space which diminishes the power output; the other is the remarkable extent to which a supercritical fluid expands when heated. (The gas which yields work in a Stirling engine is that whose expansion is shifted from one cylinder to the other; any other space – typically that between the cylinders – has to be heated and cooled but is not converted into work and thence detracts from the functioning of the engine. Because a liquid is incompressible, that space is no problem in a liquid engine.) And a supercritical fluid does expand astonishingly.

---

1   Robert Sier, *J. F. J. Malone*, p.30.

The study of supercritical fluids goes back to 1822, when Baron Charles Cagniard de la Tour discovered that if he rolled a ball down the barrel of a cannon which he had filled with fluid, stoppered tightly and heated on a fire, the sound the ball made would vary with temperature, pressure and the nature of the fluid. Both the physical and the chemical properties of fluids change at supercriticality and over the succeeding century this knowledge was used for a surprising variety of purposes. It is still a very lively field and to it we owe many perfumes as well as decaffein-ated coffee and some developments in microelectronics.

Malone was a prolific inventor: from 1913 to 1957 he would patent literally dozens of devices, in the UK, USA, France, Germany, Austria and Switzerland. Most of these were for instruments, control devices, gauges, pumps, compressors, etc. but some of them were for engines: internal and external combustion engines and improve-ments upon them. The first two patents, lodged in 1919 and 1920, while unusual for the time in proposing to use solar radiation, are not of direct relevance to our subject. But the application lodged in 1922 is: An Improved Heat Engine Operated by the Force of Expansion of a Liquid. This is, in principle, a Stirling engine. It consists of a long, thin cylinder in which a piston works; the closed end is kept hot and the other, through which the piston rod passes by way of a gland, is kept cold by running water. A displacer is a sliding fit on the piston rod and consists of wires or tubes. Its business is to shunt the working fluid from one end of the cylinder to the other, while accepting and rejecting heat. The working fluid is a 'liquid of high thermal conductivity capable of maintaining one physical state throughout its entire thermal range'.[2] Malone tells us that the temperature difference between hot and cold ends may be 750°F and the pressure ranges from 50 to 18,000 psi – so he appears to be thinking in terms of a working fluid in supercritical condition.

The following year, 1923, Malone applied for a patent for An Improved Heat Engine Operated by the Force of Expansion of a Liquid. This is a much more practical-looking machine, and he tells us that one had actually been built and tested. There are three cylinders, in each of which the displacer was 75 inches long and had a stroke of 5 inches. The temperature range was from 90°F to 650°F.[3] In 1924 yet another improvement was patented: this time for changes in the displacer or regenerator. All the above seem to have been fairly small-scale engines, whose early versions he tried using different liquids as medium. Mercury appears to have met with his approval, but was discarded on grounds of cost; thereafter water was the operating medium.

His most productive year seems to have been 1924, for all the main applications for heat engines were made that year. One thing strikes the reader as curious in the accounts of the engines and the patent specifications: in all of these, Malone says quite clearly that his engine is not hot. It is hard to see how this squares with the declared operating temperatures.

In 1931 Malone read a paper before a meeting of the Royal Society of Arts, in which he described his engines. In this paper he told his audience that he had tested a

---

2   UK patent GB202337.
3   UK Patent GB216574.

number of different working mediums such as oils, spirits, mercury, and liquid gases such as carbon dioxide and sulphur dioxide, but had concluded that water was probably the best – and certainly the most easily accessed – medium. He discusses different aspects of the engine at length and with obvious expertise. Most of the remarks appear to refer to the large engine which was built in 1925. This is a mighty device, massively constructed and standing about 20 feet high, with a 7-foot diameter cast-iron flywheel which must have weighed 10 tons. Malone describes the regenerator in detail: a multi-tube affair in which the water is shifted back and forth, acquiring and losing heat as it does so. He calls it the 'TD pile' and says that it has run continuously, driven by a cam, for 20,000,000 revolutions without any attention or maintenance. The engine had eighty large regenerator tubes and two engine cylinders. There is no mention of the theory of heat upon which Malone's reasoning is based, probably because most of that reasoning is of a baldly empirical sort which does not require much of a base in theory.

The large engine was built as a demonstration of the principle of the liquid engine and, given its size, apparently aimed at convincing Malone's most important potential customers: the builders of ships' engines. It even looks not unlike a big marine engine of the period – or, rather, like a marine engine of a previous generation, for by that time, toward the end of their reign, reciprocating steam engines were highly developed and looked rather elegant. (To an engineer, that is.)

Shipbuilders required two things of a ship's engine: one, obviously, that it would work well and reliably but, granted that, their main concern was as regards fuel consumption. Coal was still the commonest fuel and, besides being expensive, it took up a lot of room which might otherwise have been used to carry revenue-yielding cargo. Malone maintained that the results of trials by three independent engineers had shown an indicated thermal efficiency for the engine of 27 per cent. After allowing for losses in the furnace and the mechanism, this would translate into the energy delivered to the ship's propeller being 20 per cent of the energy contained in the coal. Since the relevant figures for existing marine steam engines were around 15 per cent and for railway locomotives 8 per cent, Malone's engine should have been a runaway commercial success.

That it was not, is not in doubt, for he appears to have had no orders. By 1931, shortly after the RSA lecture, Malone abandoned development of his engine and turned his attention to a heat engine which used a gas medium. It seems likely that he had little choice: ten years of continuous development work must have consumed a great deal of money and there are indications that by then he was running out of cash. The failure of the engine to attract customers is not hard to explain. By 1931, shipbuilding in north-east England had collapsed; the Western word had entered a trade depression which would not lift until the 1939–45 war brought government stimulation of the economy on a gigantic scale, by which time Malone's engine had been all but forgotten. And the market had moved on: big marine engine-builders were turning to steam turbines and reciprocating engines were coming to be regarded as things of the past.

It is odd, though, that a novel first mover of demonstrated worth should just have been ignored – the depression notwithstanding. It is difficult to avoid the impression that Malone's attitudes may have contributed. While not exactly a lone inventor, he clearly did not see himself as part of a community of likeminded professionals. His choice of the Royal Society of Arts for his public debut is perhaps significant: an organisation unaccustomed to the kind of technical presentation which an engineer would make to fellow professionals. Why not the Institute of Mechanical Engineers, or one of a number of similar bodies? The answer to this is probably to be found in Malone's deliberate distance from fellow workers in the field, his suspicion and resentment. It is a pity, for recent modelling of his engines has indicated that his claims for the performance of his large engine were probably close to the mark – though the presumption of further improvement would probably not have been justified. Malone's work would be forgotten for more than half a century and, though his ideas were to be revived, it would not be in the context of large marine engines.

CHAPTER 7

The Giant Killers

When she was commissioned in July 2003, the USS *Ronald Reagan* was the most expensive warship ever built: she cost $6.2 billion. A Nimitz-class aircraft carrier, she is over 1,000 feet long and weighs more than 100,000 tons. To work her, she carries more than 6,000 people. She is propelled by two Westinghouse nuclear reactors driving four turbines which turn four propeller shafts, delivering over quarter of a million shaft horsepower.

She carries ninety fixed-wing aircraft and helicopters. Of the former, an inventory in 2004 showed a squadron of Super Hornets (cost $70 million each), a squadron of Hawkeyes ($176 million each), a squadron of Seahawks ($43 million each) and a contingent of Greyhounds ($40 million apiece). What armament these things carry we do not know – not that it matters, for its destructive power is beyond imagining. What we do know is that the destructive power of US aircraft carriers is such that the US Navy is the second most powerful air force in the world – second only to the US Air Force itself.

She is equipped with two different types of air search radar, fire control radar, two air-traffic control radars, instrument landing system radar, three NSSM guidance systems and three Mk 95 radars. For electronic warfare and decoys she employs a Countermeasures suite and against torpedoes she has a Nixie Countermeasures suite. Her armament (besides that carried by her aircraft), includes Evolved Sea Sparrow missiles, Rolling Airframe missiles and a Close-in weapons system.

The USS *Ronald Reagan* is a colossus, a leviathan, conceived on a scale which an ordinary human being finds hard to comprehend. She is a floating fortress and there are few nation states whose total armament compares with hers. She is as nearly invulnerable as a ship can be. Only two things have been found to which she is vulnerable: small Swedish submarines and jellyfish.

None of this information is classified – you can find it all online. There you will also find the ship's service history, from her commissioning date in 2004 onward – and an honourable history it is, too. One thing is noticeable, though: there is no mention of what she was doing in 2005, the year following her commissioning. We know that that year she took part in war games off her home port of San Diego, as part of Carrier Strike Group Fifteen. It is possible that the omission of any account of her doings that year is not to do with military secrecy but is down to nothing more than simple embarrassment.

By 2005, the US Navy had retired all its small submarines, its strategists reasoning that the fleet of nuclear-powered and nuclear-armed boats would do all that submarines *could* do. But someone in the navy must have had doubts, for the Swedish Navy was invited to send one of its little Gotland-class boats to take part in the manoeuvres off San Diego. (It is possible that the doubts emanated from reports of the Gotland's exploits in the Mediterranean a few years before. The Gotland and her two sister ships had taken part in multinational training exercises in which they were undetected by submarine-hunting enemy ships which they were actually watching at the time. They had run rings around Spanish, French and American warships.) In naval as in terrestrial war games, there is a defending and an attacking force. For the exercise off San Diego, the USS *Ronald Reagan* was at the centre of the carrier strike group and circled by screens of fast-attack nuclear submarines, destroyers, cruisers and frigates, plus innumerable smaller craft, all armed with defensive detection systems – and scattered over a hundred miles of sea. No enemy could conceivably penetrate through all that to the Red Zone where lay the great carrier. One commentator described it as 'literally the most defended and heavily surveyed area in the world'.

Nuclear-powered super-carriers are unlike most ships in most ways but they *are ships*, which means that to continue to operate they must stay afloat. If you knock a big hole in them, they will sink and their usefulness will be at an end and their crew drowned. These things are elementary but appear occasionally to get forgotten in the case of ships as big as the USS *Ronald Reagan*. When the results of the war game were examined, it was found that the Gotland had evaded all the anti-submarine defences, penetrated all the screens, sunk a number of the defending vessels and torpedoed the *Ronald Reagan* – not once but several times, having made more than one attack – and had got away without anyone at all being aware that she had even been there. And, had this been for real, the USS *Ronald Reagan*, together with her 6,000 crew and all her aircraft, would have been at the bottom of the sea. No doubt the US Navy Board thanked their lucky stars that the United States was not at war with Sweden. Of course there was an enquiry, conducted at a length of which only a vast military bureaucracy was capable. The enquiry asked many questions, but there were only two that mattered: what was this Gotland submarine and why had she not been detected?

The Gotland boats had been commissioned not long before. In the years after 1945 Sweden, which had maintained a precarious neutrality in the Second World War, found itself again in an uncomfortable position. Having an eye on its geographical proximity to one side in the Cold War – Russia – it decided its best course would be to continue its long-established tradition of siding with neither of the great power alliances. Sweden's ability to maintain that tradition is partly because of its geographical situation, and partly through its minute but ferocious military capability. In strategic terms, the latter is what might be called the weasel-and-fox strategy: the weasel is very small, and both he and the fox know that the fox can easily kill the weasel, but the fox knows that in killing the weasel he is likely to be badly bitten, and so, unless he has a very good reason to do otherwise he leaves the weasel alone. Sweden didn't join NATO but neither did she encourage overtures by the Soviet Union – and she got on with building her already advanced industrial and technological base.

In keeping with the strategy, the Swedish Navy

furnished itself with vessels capable of defending its home waters: both surface vessels and submarines. In the Cold War, Sweden was a threat to neither of the great powers but was rather closer to the NATO alliance than to the Soviets, not least because, under a policy of socialised capitalism, the country was becoming prosperous. For many years the Americans were suspicious – they said, if you're not with us, you must be agin us – but eventually they mellowed to the extent that they regarded the Swedes as friendlies and even occasionally invited them to join in NATO naval manoeuvres.

The Swedes were realistic about what their navy was intended to do: it had to defend the country against any incursion into home waters. Ultimately it would be unable to defeat an invasion force from the most likely invader, the Soviet Union, but it could make any such invasion sufficiently costly to deter all but a major assault. The Baltic Sea, in which it was most likely to operate, is shallow and the waters around Sweden shallower still. Besides surface ships, there were good reasons to build a small force of small but potent submarines. And fortunately Sweden possessed a highly advanced industrial base, so that they were able to build their own boats and did not have to depend on hand-me-downs from superior powers.

The Gotland-class boats were tiny by the standards of the US Navy: only 1,600 tons displacement, 200 feet long and cheap (again by US criteria) at $100 million. They might not have been big or dear, but they were very, very clever. Their propulsion was diesel-electric, which was standard for small subs at the time, but they had something which no other submarine had: AIP. The letters stand for Air Independent Propulsion, and it allowed the subs to stay submerged for weeks at a time. Until the Swedes' invention of AIP, submarines (other than nuclear-powered) would stay on the surface or at periscope depth while a diesel engine charged the batteries, and the submarine's range in that configuration was indefinite. When submerged, the boat was propelled by electric motors which drew their power from her batteries – which gravely limited her effective range. In an attack, the sub would normally run at periscope depth until within striking distance, whereupon she would submerge and run on batteries.

In calculating the distance to be run on battery power, she would have to reckon on twice the distance from the target since, unless it were a suicide mission, she would want to get away as well as approach. The reason for the screen around the USS *Ronald Reagan* was to ensure that any attacking sub would have to submerge long before she was within reach of the Red Zone, even if she was on a suicide mission and didn't have to think about her escape. And because even the best-run diesels are noisy brutes, such a sub would be detected long before she got within effective range. Yet the Gotland sub had never been heard at all by anyone in the defending force.

The reason, of course, was AIP, which allowed the Gotland to approach submerged and undetected until she was within striking distance of her target. As long as the sub was underwater, the only way the defenders could detect her was by the sound she made. Their sonar detectors were very sophisticated: they could locate a diesel many miles away; they could even locate a nuclear submarine – for nuclear reactors need a lot of cooling water, which means pumps and the relatively quiet pumps could be heard. The AIP, on the other hand, was perfectly silent. Its engine burns diesel fuel using liquid oxygen but it makes no noise because it is a Stirling engine. On her Stirling engine, the Gotland boat can stay submerged while running silently for two weeks – ample time to approach even a US Navy Red Zone undetected and to make her escape.

Once they had recovered from the shock, the brass hats of the US Navy made a commendably rational response: they hired the Gotland for a year so that they might learn from her. She is a very high-tech boat: she mounts twenty-seven electromagnets whose purpose is to make her magnetic signature opaque to magnetic anomaly detectors; her interior machinery is mounted on rubber pads and coated with rubber acoustic-deafening buffers to prevent detection by sonar; her hull is coated with sonar-invisible material. Her external design makes her very manoeuvrable and her offensive capability is quite prodigious for a craft of her size: she has a combat-management system which allows her to guide torpedoes to multiple simultaneous targets. But the core of her invisibility is the Stirling engine. The US would ask for another year once the first one was up. There is no suggestion they were taking the Gotland apart and filching technology, merely that they were developing counter-measures which would allow them to cope with such a craft. It was well that they did so. In subsequent years, Stirling engines would be used to power submarines by the navies of France, Spain, Japan, Germany, Pakistan, Israel, Russia and South Korea. No doubt the Swedes made a healthy profit from licensing the technology.

This, without doubt, marks the maturity of the kinematic Stirling engine. It has found a niche to which it is suited better than any competitor – and, especially, better than any internal combustion engine. The only sad thing about the situation is that we don't get to know anything about it, for all such information is classified at the highest level. What we can do, though, is trace how the engines got there.

—

There are three matters still hanging in the air. One is the matter of the jellyfish. In 2006 the USS *Ronald Reagan* paid a visit to Brisbane. While moored in Brisbane harbour, the cooling water intakes ingested 860 kilograms of jellyfish. This, as you may imagine, did them no good at all (neither intakes nor jellyfish). The main reactors on such a vessel cannot be shut down during short visits, so the lack of cooling water was a serious problem. Quite how the jellyfish were removed has not been disclosed but, curiously, the weight of jellyfish was not considered a security matter so at least we know that much. It would be interesting to know whether anti-jellyfish technology was devised – and, if so, what it consisted of. Thousands of boat owners would love to be told.

The second matter is about the sinking (however theoretical) of the USS *Ronald Reagan*. When the great American press got hold of the story, they splashed it, with all the gory details. (The absence of actual gore did not seem to deter them: there was potential gore, which was enough.) But the most interesting aspect of the splenetic coverage was surprising (to me, at least): what really got under the skin of the American scribblers was the fact that a $6.2 billion carrier might have been sunk by a little submarine made by people nobody had heard of which cost a measly hundred million dollars – about the same as one of the F-35 fighter-bombers on her deck at the time. *That* was what hurt.

The third matter is, how the Stirling engine got from its low point in the early 1930s to a pinnacle of perfection sufficient to attract first the ingenious Swedes and thereafter the navies of half the world. Again, because such things are classified, we cannot tell the latest part of the story. But we can trace the earlier part, as follows.

(And there is a fourth matter which nobody remarked upon at the time – or after it, that I'm aware of. This is, that at the core of the whole affair which tumbled the mighty USS *Ronald Reagan* was an engine which had been invented two hundred years before. There was lots of mention of Ronald Reagan but none of Robert Stirling. Nobody who knows anything at all about both of these people can be in much doubt about who was the more admirable of the two.)

—

The Dark Ages of the Stirling engine lasted until nearly the end of the 1930s. Change, when it came, came in an unlikely place. At Eindhoven in Holland there was a company called Philips which sold, among other things, radio sets. The company had been formed in 1891 by Gerard and Frederik Philips, a father-and-son firm which began by making carbon-filament lamps and by the 1930s had diversified into numerous other electrical products. Among the latter were radios: very advanced by the standards of the time, the radio sets had built-in loud-speakers and came in stylishly designed art-deco cabinets. Using superhet valves, the radios were power-hungry and required a substantial electricity supply. Not satisfied with building radio receivers, the company had in 1927 set up its own short-wave radio station.

Having been early colonists, the Dutch had listeners on the other side of the world and in many places where electricity supply was uncertain or unavailable. The obvious course, for a manufacturing concern, was to supply a small portable generator to go with the radio set. Sometime in the early 1930s, the company debated how such a generator could be powered. Reliable, small, clean, two-stroke petrol engines were a couple of generations and a world war away from realisation, so the company's engineers searched for a suitable first mover which would be both trouble-free and foolproof. Among the alternatives (of which there were not many), they hit on the Stirling engine, an almost forgotten technology which had had no serious development for two generations.

Professor Gilles Holst, who by then was in charge of the Physical Research Laboratories at Eindhoven, was deeply impressed by the discrepancy between the theoretical thermal efficiency of which the engine ought to be capable (50 per cent) and the figure for actual engines (about 1 per cent). He determined upon a research programme to discover how the real efficiency of the engine might be improved. A team of researchers was established, under the direction of H. Rinia, who would later become renowned among Stirling enthusiasts for his innovative engine designs. The research was not secret – any more than any ordinary commercially funded research would be – but when, in 1940, Holland was overrun by the Nazi invasion, it was kept as quiet as possible. (Since the engine was largely opaque to most outsiders this was not too difficult.) The work proceeded throughout the Occupation without too many problems, but by 1944 the SS had got wind of a secret new engine which was being developed. Since at that time the tide of war had turned against Germany, a search was on for any technological development which might reverse the trend. A Spreng Kommando raid yielded examples of the suspected new engine – which were incomprehensible – and cylinders full of its secret fuel, which turned out to be nothing more than compressed air.

The work continued and by 1946 the results were published in the first of three articles in the *Philips Technical Review* which announced the rebirth of the Stirling engine. The new engines, compared with their predecessors, were fifty times more powerful per unit of weight and ten times faster. They were small, quiet and reliable. Two types were presented: a single-acting, single-cylinder model almost identical to Stirling's 1816 patent, and a double-acting, Rinia design, both having pressurised crankcase gas. One of the small engines was run continuously for 2,000 hours, after which it showed little or no wear in the working parts. The small engine would become the prototype of the 102-series quarter-horsepower generators which by 1953 would become more widely available.

Alas – and this was becoming a common refrain – by the early 1950s transistor-based radio sets were becoming common. So were smaller, longer-lasting, cheaper batteries. Transistor radios worked on much smaller power inputs than valve radios and the combination of the two factors killed the market for the small Philips generator. Production ceased in 1954.

Research into the larger, Rinia-designed engine continued, but it was plagued by sealing problems, principally as regards the piston. Because the working gas was common to crankcase and cylinder, lubrication created difficulty, for most lubricants would degrade and pollute the working fluid. Without adequate lubrication, the piston would cause wear in the cylinder, leading to more sealing problems. (In a single-cylinder engine, the connecting rod lies at an angle to the cylinder save when the piston is at top or bottom dead centre. As the piston rises and falls, there is a sideways thrust which causes uneven wear in the cylinder, leading to loss of pressure through the seal. This is not serious in the case of a small engine, but as size increases, the effect becomes more acute. It is yet another of the scaling problems we have encountered in earlier engines.)

By the early 1950s this had come to seem an insuperable problem – but it was one which would be solved by relatively simple mechanical means: an arrangement of the cranks known as a rhombic drive. This is difficult to describe but quite simple in practice. The single-cylinder engine has two crankshafts instead of one. Piston and displacer rods are each fixed to a yoke which is connected to the crank by a bridle and the four bridles form a rhomboid. The cranks are geared to each other and rotate in opposite directions, making the rhomboid expand and contract, causing piston and displacer to rise and fall. (All this is difficult to visualise: the best course is to look up 'rhombic drive' online, where you should easily find an animated drawing.) The benefits of the rhombic drive for the Stirling engine are twofold: the crankcase gas can be separated from the working gas (which allows for higher pressures with few sealing problems) and, more importantly, the piston and displacer rods are perfectly in line with the cylinder, thus eliminating the troublesome wear and sealing problems.

Thus endowed, the Rinia-design Stirling engine looked set for a bright future. Air as the working fluid was pressurised, which gave an increase in power; then hydrogen and helium were substituted for air, at higher and higher pressures. Each increase augmented the efficiency of the engine, though at the expense of a recurrence of sealing problems. (Hydrogen and helium are very small molecules and it is notoriously difficult to contain them, since they are often smaller than the spaces between the molecules of the containing material.) But by the later 1950s thermal efficiencies of 38 per cent were being yielded by rhombic drive Stirling engines. This was better than petrol engines and than most diesels. In 1958, General Motors negotiated a licence with Philips, with a view to developing Stirling engines for use in propelling various vehicles, as well as in providing electricity for military use – and, tentatively, for extra-terrestrial use. In the latter they were a trifle premature but, such was the Americans' annoyance at the Russians' exploits, they were prepared to contemplate just about anything.

General Motors put a huge effort into developing Stirling technology in all sorts of applications. They even fitted out the first Stirling-engined car, the Stir-Lec 1. This used a Stirling engine to power a battery-bank, which provided the drive through an electric motor – a concept not a million miles away from currently fashionable proposals for a hydrogen-powered vehicle. There was a solar-powered generator for use in space, which was reminiscent of Ericsson's. All looked fair for the future when, about 1960, a decision was taken at some high and opaque level in the General Motors Corporation that this was all a diversion from their main task, which was, of course, making money, not engines. A lot of people lost jobs they had built their careers around and a whole lot of promising research disappeared into the grave of commercial confidentiality, from which it was never to return. The one small upside of the affair was that it would provide a happy hunting ground for a generation of conspiracy theorists.

Through the 1960s there was a great deal of activity around Stirling engines, in the US and in Europe. A substantial literature grew up and the advent of computers enabled a level of analysis which previously would have been unthinkable. With the proliferation of internal combustion engines, atmospheric pollution had become a matter of serious concern for public health authorities around the world. Since the heating of the Stirling engine could be more easily controlled than the burning of fuel in an internal combustion engine, that produced an incentive to the engine's development. By the end of the decade, Philips had a four-cylinder rhombic-drive engine of 200 horsepower which was capable of powering a bus. Both MAN and the Swedish firm United Stirling took out licences to produce and develop the engine. As has happened so often in the history of Stirling engines, improvements in materials brought old designs into the realm of the possible. The Rinia engine which had taken a back seat a decade before was revived using new self-lubricating seals – with the result that the Ford Motor Corporation entered into a joint venture with Philips to develop an engine for the 1980s.

The main driver of Stirling automobile research in the USA was the Arab oil crisis of 1973. Suddenly and alarmingly, there was the prospect of an oil shortage. To a country such as the US, whose whole culture was based on an assumption of cheap and abundant fuel, this was shocking. Directives went out to branches of government to do something about it. As has commonly been the case in the USA, ideological restrictions on government involvement in industrial research were cast aside or disguised as military spending, and the best-qualified organisations were awarded the contracts. There could be no real question as to who the latter were: NASA had the capability and the organisational structures. Besides, they subcontracted work where possible to major corporations, so everyone got a bite of the cherry.

It happened that NASA already had an interest in Stirling technology. Their Lewis research organisation had made a start, with a view to its possible use in space-related projects. The Lewis branch had a contract with

Ford for research on a Stirling unit for automotive use which, provided by Philips, ran for a little over a year until Ford decided to pull out. Lewis, having changed its name to Glenn, then transferred the work to companies which had a track record in the field: Mechanical Technology Inc., of Albany, NY, with United Stirling of Sweden as a subcontractor. United Stirling were licensees of Philips in the Netherlands, with rights to use and develop their designs. The enterprise was highly successful, and new and improved automotive Stirling engines were built, mostly using hydrogen as the working fluid.

NASA formed a partnership with the US Department of Energy to build further Stirling-powered vehicles. A Spirit saloon from the American Motor Corporation was engined with a unit from United Stirling. In road testing, this vehicle covered more than 50,000 miles at an average fuel consumption of 8.3L/100km – a respectable but not spectacular performance. A VAM Lerma hatchback using the same engine also turned in a respectable performance.

AMC models Spirit and Concord, as well as an Opel, were also converted to Stirling power: all worked well, though, like the others, they suffered from the Stirling engine's built-in defects: there was an appreciable delay on first using the vehicles, as the engine heated up to working temperature, and it was difficult to alter the engine's speed: the latter a serious defect in the eyes of a driving public accustomed to the instantaneous response which by then even diesel-engined vehicles would provide. Research showed it would be possible to overcome both deficiencies: a cylinder-head electric heater powered by the car's battery could eliminate most of the start-up delay, and an improved drivetrain would help with acceleration. By 1985, the Stirling engine mounted in a Chevrolet Celebrity had yielded the highest thermal efficiency ever registered by an automotive engine: 38.5 per cent.

Motorcars, trucks and buses were powered by the engines, to everyone's satisfaction, and the new day of the Stirling engine seemed to have dawned. But, as you may have guessed from previous accounts, it was a false dawn. By the mid-1980s the panic about oil prices had died away. The price of oil, while still high by historical standards, had been whittled down by the inflation its rise had caused; new sources had been developed which were not vulnerable to Arab politics; gas-guzzling had become unfashionable in the US; and the need for economical modes of propulsion had faded. American engine makers pointed out that, although the engines undoubtedly worked, the improvements over existing internal combustion engines were not sufficient to justify redesigning vehicles and re-tooling production lines. The US government lost interest, NASA lost funding, and the engine lost sponsors. Jan Meijer was for many years in charge of the Philips engine development programme. When asked what was wrong with Stirling engines, he is reported to have said, 'The existence of other engines.'

The pattern has now become familiar: the Stirling engine is brought, with great difficulty, to a pitch of development at which it will serve an identified need, only to be supplanted by something which meets that need better. Or does it more cheaply or more conveniently or requires less investment and institutional upheaval. There were two substantial exceptions to this. The greater we will deal with in the next chapter. The smaller – and it wasn't so very small – we will look at here. It concerns the matter of Swedish submarines.

In 1983 United Stirling of Malmö in Sweden put out a press release in which it announced the summary conclusion of a recent conference: that the 4-95 and 4-275 engines had been successfully adapted to run underwater. They would run on diesel oil and oxygen and their purpose was evidently for use in a submarine, though the release didn't say so in so many words. The engines were for use at the depths at which Swedish submarines operate – the shallow water of the Baltic Sea and parts adjacent. And the exhaust would be absorbed by the seawater, so there wouldn't be any bubbles as a giveaway. The exhaust would also be mixed with the oxygen before it combined with the fuel, which would allow the temperature of the combustion to be controlled. (This was necessary, for at high temperature in the presence of pure oxygen, steel simply burns.)

The Gotland-class submarines were built by Kockums, a subsidiary of Saab, in Malmö. Kockums had licensed the Stirling technology from Philips in the 1960s and had developed the 4-275 engine to a point at which it would run almost indefinitely without maintenance. It ran on diesel oil burnt with a mixture of oxygen and exhaust fumes at a hot-end temperature of 750 degrees Celsius. Using helium as a working fluid, it would return a thermal efficiency of 39 per cent and provide 75 kilowatts of charge to the batteries. Several hundred of the engines were built over a period of ten years or so.

The first installation of the engine in a submarine took place in 1988, when a Nacken-class boat was cut in half and an 8-metre AIP section inserted. The converted vessel proving satisfactory, the new Gotland-class boats were fitted with the AIP system. The AIP was not the main propulsion: that came from two sets of conventional diesel engines which could run when the boat was on the surface or near it. Subsurface, the Stirling-engined AIP took over. The system is so successful that the Swedes have used it in their latest Blekinge-class submarines.

So: a kinematic Stirling engine which has been a big success, doing something that no other engine can do. The only pity is, because that something is military, the details of the technology are secret: a curious irony. And, of course, because the development is both high-tech and secret, any spin-offs are likely to remain under wraps. Is this the last fling of kinematic Stirling? It's hard to say – and anyway doesn't matter too much, for the most important iteration of the engine had taken a completely different direction, which is the subject of our next chapter.

CHAPTER 8

William Beale and the Free-Piston Stirling Engine

One evening in 1964, William T. Beale was pondering the Stirling engine. He had decided to prepare a seminar for the 'Special Topics in Mechanical Engineering' course which he gave as a class for advanced students in the Mechanical Engineering syllabus at Ohio University in Athens, Ohio. The Mech. E. syllabus, being crowded as such things are, allocated little time to the largely forgotten Stirling engine. It was a curiosity, no more, though it had intrigued many generations of engineers with its sheer ingenuity, its apparent promise and the inability of so many engineers to realise that promise. It intrigued William T. Beale. Professor Beale was wondering how best to describe the engine. He would tell his students how the engine worked by the alternate expansion and contraction of its working fluid, and how its little-understood regenerator contributed to its overall efficiency: how the latter in theory approached the Carnot maximum, but in practice was poor. He then proceeded to the explanation of why this was so: how problems with seals and linkages cumulatively diminished the theoretical perfection so greatly that in the real world the engine was unable to compete with the internal combustion engine.

The main source of trouble was the linkage which connected the piston and displacer to the crankshaft – and thence to the world of work. One imagines that the drawing he was looking at as he worked was the single-cylinder 1816 patent, or something like it, for that is where the linkage appears most burdensome. Professor Beale had the novel idea of asking the students to conduct a thought experiment, a procedure hallowed in the annals of science, most famously by Albert Einstein. The students should ask themselves an apparently absurd question: 'If you took a hacksaw and cut off the crank and crankcase assembly, why would that stop the engine working?' The premise on which the question was based was, of course, that the hacksaw job *would* stop the engine working, no question. It was obvious: if the linkage were removed, there would be nothing to determine the relative movement of piston and displacer, and hence all movement would cease. But, thinking idly about the question and its answer, Beale saw that the premise was false, for without the apparatus of cranks and connecting rods, the thing could be made to work. It would need some springs and things, but it should be possible.

Beale seems to have experienced a moment of intense insight, the sort of thing that in religious circles would be described as an epiphany: he saw – just *saw* – that if the linkages were to be removed, the engine could still work. It would be able to go on running without them. And in that moment of insight the free-piston Stirling engine was born. Born but not raised: the raising of it would occupy the rest of Beale's life. Not just his working life, for from then on it would be the background to his whole existence. At the time of this perception he could not have imagined just how difficult was the task allotted to him, but he had what few of us are accorded: a clear vision of his purpose in life. In his case, to realise the free-piston Stirling engine.

(From now on we shall call him William, rather than Professor Beale, just as we refer to the Reverend Dr Stirling as Robert, for we like our heroes to feel familiar. Since neither of them was the sort of person to stand on ceremony, the presumption may be permissible.)

William Beale was the best sort of American engineer: a quiet American engineer in a country whose pre-eminence since the early days has derived largely from the excellence of its engineers – a matter which seems to be in danger of being forgotten in these dark days. He was born in 1928, in Chattanooga, Tennessee, into a decade of depression and poverty for many. His father had built the family house with his own hands but, as so often happened, had given it as security for a bank loan. When, in 1936, the bank foreclosed, the only option for the whole family was to load themselves – two adults and four children – and what goods they could carry into the family sedan and head south, looking for work.

In a Great American Saga worthy of *The Grapes of Wrath*, they took work where they could find it, fortunate that David Beale was a versatile and enterprising character, as well as an experienced construction engineer. He found work building post offices, and the family finances improved. But they were still itinerants: when one post office was done, the whole family would move on to where David would build another one.

The Beales were more fortunate than Steinbeck's Joads, for by the time they reached Louisiana, the US was preparing to enter the Second World War and the economy would be transformed by an immense injection of public funds to put it on a war footing. Having attended a succession of schools, William's formal education must have been erratic, but early on he demonstrated the affinity for and understanding of machines which was to characterise the rest of his life. By the time William was in his teens, his father had bought a small farm and set up a business brewing and selling beer. Stories of his being a bootlegger are untrue, but he did use a small motor boat on the Mississippi as a retail outlet, which enabled him to evade taxes – a matter which his customers, with memories of the recently ended Prohibition, no doubt appreciated. The boat engine was William's sole charge. Since cash was short and tools expensive, he developed a canon of economy which was to be one of the criteria by which he would order the rest of his life.

As the war was ending, William became old enough to be accepted as a volunteer by the US Navy. The navy by that time had expanded to an extent inconceivable to people who had known it pre-war. Its intake system, honed by the urgency of warfare, became the New Recruit Placement Process, designed to maximise the value of recruits to the navy by ensuring that they were given work commensurate with their ability. This depended on an assessment procedure – in which William scored so highly that he was trained to operate some of the most advanced technical equipment then in use. He seems to have been so favourably impressed by the efficiency of the whole business that it would confirm him in his canon of economy. And economy took a moral colouring, in the sense that William bitterly opposed as waste any unnecessary human suffering. As the man who relayed messages from the flight deck of an aircraft carrier, he saw people killed through miscued landing reports and traffic jams: in his own words, 'The bloody waste of war impressed me permanently.'

The navy must have provided more education than mere technical training, for in the mid-'50s William was to go to Washington State College, from which he gained a Bachelor of Science degree under the mentoring of Mechanical Engineering professor Harry Sorensen. (Sorensen's take on mechanical engineering would extend to various works on heat, the study and utilisation of which would figure so largely in William's career.) From there he went to CalTech, where he got a Master's degree, and then to one of the summits of most scientists' ambition, the Massachusetts Institute of Technology, for a PhD. For bright young engineers of the time, an MIT PhD was a passport to a lucrative career in industry: in glee they would chant 'MIT, PhD, M.O.N.E.Y.' It was indicative of William's character, then and later, that this would be anathema to him. He did the unheard-of and dropped out of the PhD course.

That didn't mean that William had no ambitions as regards scientific research: merely that the road he would take would not be that of the climber on corporate ladders. He was fortunate that at the time the US

161

economy was booming, so there was no shortage of jobs for a well-qualified engineer. He seems also to have been fortunate in meeting and marrying Carol, who would stick with him through a life which in many ways cannot have been easy. William got a job at Boston University, teaching mechanical engineering. It is said that over time he taught every course in the Mech. E. curriculum. By the mid-'60s, William had become a tenured member of staff at Ohio University, where he garnered grants to set up and run a research group looking at renewable energy power plants. At some point along the way, probably quite early, he had become imbued with the second of his principal principles: a belief that human progress could best – possibly only – be furthered by the use of the scientific method. Observation, hypothesis, experiment and critical assessment were to be his watchwords thereafter. And empiricism of the purest sort, for, in engineering, theories had to pass the test of practicality: 'No ideas but in things,' William would say.[1]

There is a distinction to be drawn between scientists and engineers – though many engineers are also scientists of distinction, and some scientists are respectable engineers. Though the two often tend to merge into one another, there is an essential difference: scientists are concerned principally with knowing things, and engineers with doing things. The Ancient Greeks called it the difference between *logos* and *techne* – and the people who made the distinction, being themselves philosophers and on the *logos* side, looked down on the mere technicians.

---

1   Shelton, *The Next Great Thing*, p. 104.

In this they were in error, for knowing *how* is as valuable a mode of knowledge as knowing *that*. ('The term 'technology', which brings the two together, dates back to the early seventeenth century, though it became popular only in the 1960s.) Nonetheless, the division became traditional, though as understanding of the natural world increased, a sound grounding in science would become necessary for the most modest of engineers. The significance of William's statement above is as an indication, not of his ability to theorise (which was, by all accounts, formidable) but of how he valued theory in relation to practice.

William was, above all, a doer and a maker and could be impatient of theoreticians. (It has to be admitted that there have been a great many people happy to theorise at great length about Stirling engines.) David Berchowitz, who was a close friend and for sixteen years a colleague of William's, tells the following story.

'William and I attended a particularly useless conference – an annual meeting of the brotherhood of august Stirling researchers – in Miami, Florida. After two or three days William became visibly restless, listening to the ongoing drone of how to seal crank machines. He suddenly stood up in the middle of a meeting and said, "I'm willing to bet any of you in this audience that none of you would be able to build a practical Stirling engine!"

'The Chairman, hesitatingly: "Uh, uh, William, do I take it that you are offering a wager?"

'William: "Yes! In fact I'll give ten thousand dollars to the first person who can demonstrate a practical Stirling engine."

'Audience and Chair: Dead silence.

'Chair: "Uh, uh, William: Shall we call it the Beale Prize?"

'William: "Fine with me!"

'On the way back in the plane, William was a little reflective. So I suggested that maybe it was not too late to take back the offer since it was made in a moment of passion. "You know . . . trying to seal crank engines and all . . ." William brushed my remark aside and said that he had only one concern and that was how to explain this business to Carol. That was the birth of the Beale Prize, awarded only once and won by Catherine Chagnot, who built a hundred or two hundred crank air engines in India.'[2]

While we are on the subject of theory and practice, another of David Berchowitz's stories is instructive of William's attitude to theory vis-à-vis practice. He describes a conversation at a technical meeting between himself and William:

William: 'Dah dah dah . . . centering displacer . . . (hard-to-follow explanation) . . . I've tried it: it works.'

David Berchowitz (twenty-nine years old): 'Sounds impossible!'

William: 'Impossible! We don't use those kinds of words at Sunpower! Nothing is impossible!'

DB: 'How about opening a tin can with a banana peel?'

William: 'Freeze it rock solid with liquid nitrogen.'

DB: 'I don't know if that idea would work but, like

---

2 David Berchowitz, *Some Reminiscences of Sunpower*, December 1997.

all ideas at Sunpower, you usually have to try it to make sure.'[3]

William would work on his invention for ten years while at Ohio University: improving it, refining it, simplifying it, making it *work*. Together with his students he articulated his original idea and, in the summer of 1966, for the first time, a free-piston Stirling engine ran convincingly. William would say it was, mechanically, 'very simple, used air at atmospheric pressure as a working fluid, had an elementary annular gap regenerator, and relied on close fits for the piston and displacer rod seals. It was a single-cylinder engine, positioned vertically to minimise frictional drag, and common rubber bands were used to suspend the piston and displacer against gravity . . .'[4] I like the bit about rubber bands.

Unlike Malone, William didn't hold his ideas close to himself for fear someone else might profit by them. He made them public and, in so doing, gathered a community of likeminded people who were happy to work alongside him in developing his great idea. By 1974, he was about ready to make the move to take the idea out of academia and into the great world of commerce and manufacturing. That year he set up Sunpower Inc., a limited company which would raise funds to do research on behalf of sponsors – for whom the company would design and produce prototype free-piston Stirling engines, which the sponsor might then manufacture and sell for profit. It wasn't altogether a charity, but it wasn't far off

---

3 David Berchowitz, *Some Reminiscences of Sunpower*, December 1997.
4 Walker & Senft, *Free Piston Stirling Engines*, p.129.

being one. And a small market had already been created, for from 1970 William had sold a number of demonstration models to universities, government departments and private companies interested in the technology.

William was an idealist, but he was an *American* idealist. He understood that the rest of the world was not like himself, and that if his invention were ever to bring benefit to humanity, there must be many, many people who would make it and sell it, not for the sake of the benefit to humanity but because by doing so they would make a buck. It was a marriage of principle and realism which, if not totally unique, must have been pretty close to it in the US of the 1970s. The company would be run by William, but on the basis of consensus among its people on almost every issue of any importance. The entire company met every Monday morning to discuss the past week and that to come. It wasn't a democracy, for everyone recognised that William was the boss – but a boss who took no pleasure in giving orders and who believed that, given a rational and objective examination by intelligent and unprejudiced persons, the answer to almost every problem would present itself in such a fashion that no unprejudiced person could resist the obvious conclusion. The scientific method would run the company as well as the research.

And the gods – in which William didn't believe – were favourable, in that the gods decided to have the US withdraw from its entanglement in Vietnam just about the time that Sunpower came into being. Nobody is suggesting that the gods ended the long misery of Vietnam just so that things would work out for Sunpower – that would be disproportionate, hubristic and bad theology – but it did fall out pretty well for Sunpower. It appears that a large proportion of the student population at Ohio University in Athens, Ohio, were there only to escape the draft. When the war ended, these persons went back to doing whatever they would rather have been doing had there not been the possibility of being sent to die in the jungles of Vietnam. As a result, student enrolment dropped sharply and the Mechanical Engineering department found it had workshops surplus to requirement. William rented the Mech. E. workshop and Sunpower took on the two technicians who would otherwise have been made redundant.

So came into the world a company whose business it would be to save that world from itself by means of the free-piston Stirling engine. It was called Sunpower because, as William pointed out, the most obvious use for the engine was to convert sunlight into electricity and by so doing prevent pollution, climate change, dirt and misery. After a quick calculation, he estimated that the sunlight falling on a 170-mile square patch of the Arizona desert – a mere *postage stamp*, as he said – would provide all the energy consumed each day by the whole of the United States. The resources of the company were admittedly somewhat slender for such an ambition, but William was an optimist. And he was an optimist with a mission. Unfocused optimists are ten-a-penny; optimists who know exactly what they are optimistic about are not, and so are more likely to change the world. But, alas, his optimism was fated to be unjustified as regards this application of the engine, as we shall see later.

The resources of the company were slender, considering its ambition. A collection of dilapidated buildings, some workshop machines, a few computers and some very talented people; some patents for an engine which nobody had ever heard of. And, of course, William T. Beale.

—

Over the next few years William would reap the harvest which grew from broadcasting his discoveries about the Stirling engine. There arose a community worldwide that was devoted to the development of the free-piston Stirling engine: engineers, thermodynamicists, theoreticians, tinkerers and amateurs. All of them got to know about William Beale and his engine – and a lot of them wanted to be in on the action. So when Sunpower advertised for staff, there was no shortage of applicants. The advertisements made it clear that this was no sure-fire career opportunity: pay was poor, tenure uncertain, promotion unlikely and resources, like tenure, dependent on research and development sponsors, which might or might not exist and whose support was liable to evaporate with little warning. In a buoyant economy, this sort of advert would barely produce an office cleaner; however, it attracted an extraordinary galaxy of talented engineers, scientists and technicians from all around the world. They would form a nucleus around William and over time bring into the real world the engine which until then had existed only in William's head.

The atmosphere of such a company is hard for most of us to imagine. A firm in which every employee is possessed of an acknowledged expertise; in which creative tension is encouraged but moderated by openness and discussion; in which decisions must be justified in open debate among peers. And of course, presiding over it all, the figure of William: a boss who chose not to understand deference, a director who preferred not to direct, but whose wishes were respected out of respect. And a company which collectively would retire to the local bar on every significant occasion: for celebration or commiseration.

We are fortunate in having a first-hand account of life inside the company, for the writer Mark L. Shelton spent two years in the early 1990s embedded in the company as a privileged observer. The outcome was a colloquial but elegantly written account of the daily trials of the people who made up this most unusual high-tech research and development company.[5] It is a work which is as difficult to describe as was its subject: it is a day-to-day stream in which personalities, technicalities, explanations, policies, enemies and allies are all mixed up with wit and laughter and thermodynamics and, of course, Stirling engines. Its subject, though, must not have been as isolated from institutional consciousness as its text implies, for the book was part-funded by a fellowship from the Ohio Arts Council. It is to be recommended to anyone with an interest in an extraordinary technology and, in William T. Beale, an even more extraordinary man.

By 1978 the company had completed to its own and its sponsors' satisfaction its first commercial free-piston engine, the RE-1000, which was duly tested and delivered,

---

5   Shelton, *The Next Great Thing*, p.104.

and entered the coveted commercial market. Sunpower would continue to invent and develop engines, wherever it and – more importantly – its sponsors (existing or potential) saw an opportunity. Throughout the '80s it would design, develop and test free-piston Stirling engines of powers ranging from 35 to 75 watts, fuelled by various heat sources. Among the latter were oil, natural gas, biomass and, of course, solar. Engines of 2kW power were developed and, later, a 3kW machine for the US Army. There were plans to build larger machines but the development process was expensive so ambitions had to be tailored to sponsors' requirements. Because scaling up has always given rise to problems in Stirling engines – what works well at one size does not at another – any departure from contracted work could prove ruinously expensive.

So what is a free-piston Stirling engine *like*? If you have stuck with me this far, I expect you will have some idea. But, as we all know, Stirling engines are not easy things to understand and even less easy to visualise. The free-piston Stirling engine is, without doubt, the least easy in that respect. It consists of a tube closed at either end, in which there are piston and displacer. William Beale was fond of saying that when a technician accustomed to internal combustion engines first dissembles a free-piston Stirling engine and looks at all the parts on his workbench, he will invariably start looking around under the workbench assuming that several important parts must have rolled off the bench. The technician is dumbfounded when he sees how few parts there are.

As in a traditional Stirling, the displacer moves gas from hot to cold end and back, and in the process creates a variation in pressure which moves the piston. But, since it's a *free*-piston Stirling engine, the piston isn't mechanically connected to anything. It just slides to and fro accordingly, as the gas pressure rises and falls. That, you might say, may be interesting, but because the moving piston isn't connected to anything, how can it do any work? How can it affect anything in the world outside the tube?

Being a Stirling engine, it is of course very clever, so useful work is got out of it in a very clever way, by means of a linear alternator. The alternator in your car works by rotating a magnet past coils of copper wire, so that the motion induces a current in the wire. The essence of this lies not in the rotation but in the motion per se, and a magnet passing a straight line of wire will do the same, though less conveniently. The polarity of the field will reverse with the direction of travel. In the free-piston Stirling engine the magnet is attached to the piston rod and it passes through a coil of wire which is fixed in relation to the engine casing, in which it induces an alternating current. The current can easily be got out of the tube through wires without disturbing anything and so work can be extracted from the engine.

The schematic drawing (plate 18) of a free-piston Stirling engine gives a general idea of the layout. The working gas moves between expansion space and compression space and back, and at each movement passes through a regenerator. (The regenerator is still with us, even in this highest of high-tech applications.) The increase in pressure produced by the gas being heated causes the piston to move until it is brought up against a spring, which pushes it up again. In the later versions of the Sunpower engine, the spring is a gas-spring, thus eliminating a part which might wear out. Similarly, the piston, displacer and rods are held on gas bearings whose pressure is generated by the motion of the engine itself – thus eliminating both friction and wear. The result is an engine which will run for a very long time indeed, without repair or maintenance. It doesn't look like much, but it's a very, very smart piece of kit.

The engine in the illustration looks very simple: it *is* simple, compared to earlier, kinematic versions, and compared to almost all internal combustion engines. It looks to be the sort of thing which any competent machine shop could turn out in a few days. Why, then, should a talented engineer like William Beale have to spend a large part of his life to get the thing to work properly? The reason is that the simplicity is an illusion: it obscures from view the practicality of a machine whose implementation is of staggering complexity. As Graham Walker writes in *Free Piston Stirling Engines*: 'The free-piston Stirling engine is a paradox of apparent simplicity that is, in fact, the most difficult of Stirling engines to reduce to practice.'

—

Solar-powered Stirlings were never far from William's mind, though naturally the company had to do what it could get funding for. Almost all the engines required purpose-built heat sources, so that part of the R&D effort had to go toward developing those. The basics of a solar-powered Stirling engine had been worked out by Ericsson more than a century earlier, and by the late 1970s the technology of producing an automatically orientating solar array presented no great obstacle. The free-piston Stirling engine was an obvious fit for this, so some of the Sunpower effort continually turned in that direction (the company was called *Sunpower*, after all). By about 1990 the company was in a position to demonstrate the successful application of a solar-powered engine. (An inventor named Roelf J. Meijer had demonstrated something similar in 1987, which wasn't quite so satisfactory: the idea of applied Stirling was by that time pretty widespread. NASA had followed up their earlier interest in 1976 by patenting a solar-powered pump.) Given the oil shocks and energy panics of the previous few decades, solar-powered free-piston Stirling engine generators looked like a shoo-in.

Sunpower were not alone. Besides several US companies, United Stirling in Sweden had participated with MIT in a NASA-funded study in 1978 which looked at both free-piston and kinematic generators for solar installations. In 1979 they began to develop their 4-95 Stirling engine, with a view to its employment in a solar converter. By 1980 they had branched out on their own. They tested solar energy sets at the Georgia Institute of Technology in 1981 and at the Edwards Air Force Base in 1982, following which they announced the installation of the first solar Stirling engine with a hybrid receiver in a parabolic dish. The 4-95 engine was of course a kinematic Stirling engine and therefore a much heavier piece of kit than a free-piston Stirling engine. And it had lubrication problems: since the receiver would have to follow the sun, the original sump-oil drainage would be unworkable – a problem which the free-piston Stirling engine didn't have. It was also very expensive – and a large (and therefore expensive) converter would need a very large array of mirrors, which would be vulnerable to storm damage.

Things were looking promising for solar energy as the new millennium dawned. Quite soon NASA would fund a Sunpower program to develop the next generation of Advanced Stirling Converter. In 1996 Stirling Energy Systems (SES) had been set up in Scottsdale, Arizona, with very large funding to 'develop equipment for utility-scale renewable energy power plants using parabolic dish and Stirling technology'. SES formed associations with NASA-Glenn Research Center, Kockums Submarine Systems, the US Department of Energy and the Boeing Aircraft Company: big hitters by anyone's reckoning. The object was 'to become a premier worldwide renewable energy technology company to meet the global demand for renewable electric generating technologies through the commercialization of its own Stirling cycle engine technology for solar power generation applications'.

Well: they were certainly strong on verbiage. Not so good at seeing what was coming, though. The Chinese government, which could also see a potential need for non-fossil-fuel generating capability, decided not to compete but instead to fund the development of photovoltaic panels. Being a Communist government, they were less than forthcoming when asked how much they would spend, but the subsidies were evidently in the billions. The market for Stirling solar electric generation collapsed when Chinese photovoltaic panels could be installed at a cost of about $0.30 per watt of installed electricity. On 29 September 2011 Stirling Energy Systems filed for Chapter 7 bankruptcy. One cannot but wonder what would have happened if the US government had funded the covering of a piece of Arizona with Sunpower Stirling-engined converters. And, of course, the irony: the world's biggest Communist state has captured what promises to be the biggest commercial market of the early twenty-first century because the world's biggest capitalist state was slow to organise the funding and development of an indigenous technology. That may be unfair; we shall probably never know. But the suspicion will remain that if a 170-mile square of the Arizona desert had been covered with Stirling generators, the history of the world might be different.

—

In 1989, William wrote a letter to the *New York Times*:

> If we could gather the collective will, we could supply, carbon dioxide free, all of the energy this country needs for transport, industry, domestic heating – everything from an area of our southwestern desert about 170 miles on a side. That's a conservative estimate. The cost, even neglecting the saving from not doing more expensive energy alternatives, would amount to about 15 years of our military budget. A lot of money, but compare the cost-effectiveness of a 5,000-mile-long sea wall and all the things that have to go with it.

William's ideas for saving the planet through free-piston Stirling engines extended well beyond solar power. The engine, he said, would lend itself to domestic heat and power generation. Since it would run on almost any heat source, the scope for this was huge, especially in a country such as the US, with a large rural population and a tradition of bucolic self-sufficiency. Combined heat-and-power installations were especially attractive, given that the source which would provide domestic heating might produce the household's power from the waste heat. In the late 1990s Sunpower licensed its free-piston Stirling engine technology in Europe and Asia for use in micro combined heat-and-power installations. By the year 2000, Sunpower had a free-piston Stirling-engined household cogeneration appliance patented, and by the following year the company to which it had licensed the thing had satisfactorily completed field trials. Such applications have been popular but have had to compete with fuel cell technology, which is substantially cheaper. As at the time of writing, neither system has succeeded in dominating the market.

By the time of his death in 2016 William had had to come to terms with the dominance of photovoltaic solar electricity but, typically, moved on and saw how it might be combined with a Stirling generator to achieve a carbon-free source of heat and power. He said, in a paper in which he asked his friend and long-time collaborator Izzi Urieli to present to the 2016 International Stirling Engine Conference, 'Foremost among these [opportunities for the engine] in the carbon-constrained world of the near future, is the biomass-fired home power plant which complements photovoltaics. Any fuel, with or without pre-processing, could be used to power a free-piston Stirling that automatically recharges a home battery bank when charge levels get low. If pyrolysis is used for pre-processing and the resulting biochar is returned to the soil, one has the ideal future power plant: un-interruptible, maintenance-free and carbon dioxide-negative in all sizes from domestic on up.' In this, curiously, he reflects what Malone had said when he envisaged a Britain in which there would be a liquid Stirling power plant at the end of every street. CHP hasn't yet saved the planet but it has made some progress and may yet contribute to our salvation.

William was prolific right to the end: he said that when he was hard at work on Stirling engines good ideas just floated by. His son Dan wrote to me:

Just before he died I asked him for some of his favourite Stirling ideas. He described four product ideas, namely:

Product no 1. Carbon-negative power plant.

Product no 2. Solar power conversion system that can be manufactured anywhere.

Product no 3. Non-polluting alternative to diesel engine for vehicles.

Product no 4. Water condenser for desert survival.

The first three are improvements on ideas we have already mentioned – but radical improvements. The fourth, though, is particularly appealing. Imagine, if you will, the lone traveller in a desert land. His supplies of water have run out and there is no sign of an oasis. The sun blazes down, death stares him in the face, until he remembers what is in his knapsack: a duplex free-piston Stirling engine which 'consists of a tube wherein a piston slides freely on a rod connecting two displacers. When one end is heated, the displacer at the hot end acts as a Stirling engine driving a Stirling cooler. There is no heat transfer fluid other than air, and no exotic materials or fabrication methods are used. A duplex free-piston Stirling engine small enough to be carried in a backpack could, in combination with a solar concentrator, create a ball of ice on demand wherever there is bright sunlight and thus provide fresh water.' What a guy!

—

It had been my intention in writing this to focus solely on the Stirling engine as a prime mover, but it seems churlish, in a chapter about William and Sunpower, not to mention the reverse side of the gold coin which is the Stirling engine: the cooler. The Stirling engine is a perfectly symmetrical machine: provided with a temperature gradient, it will produce work; provided with work, it will produce a temperature gradient. Put it another way: if you give it heat and take the waste heat away, it will turn a flywheel or produce electricity; if you turn its wheel or give it electricity, it will act as a heat pump, taking heat in at one end and putting it out at the other – and it will do this against an existing temperature gradient.

This had been known about since the mid-nineteenth century and was exploited in the twentieth, principally by Philips, who by the 1950s had produced a cryocooler which was capable of liquefying air without the pre-compression which is normally a necessary prelude to such operations. William had, of course, known about this and one of his aims for Stirling engine-impelled human betterment involved an efficient cooler. Throughout the 1980s, Sunpower developed and tested cryocoolers, until by the late 1990s the cryocooler had grown from its conceptual beginnings to a robust and reliable piece of engineering. In 1999 Sunpower created a separate cryocooler manufacturing operation. The free-piston Stirling engine was well-suited to cryocooler operation: its linear alternator worked well, with little alteration, as a motor.

The Stirling cryocooler will produce temperatures as low as 200 degrees below zero Centigrade (73K), at which the constituent gases of air – oxygen, nitrogen and argon – will liquefy. Two-stage Stirling coolers will lower this to around 20K, at which point neon and hydrogen become liquid. The applications of such very low temperatures are naturally somewhat specialised, but higher, though still cryogenic, levels are more common: many electronic sensors and microprocessors require such temperatures and, for attaining these, Stirling cryocoolers are the best technology available.

In 2002 NASA's RHESSI mission used a Sunpower M77 cryocooler. RHESSI is an acronym of Reuven Ramaty High Energy Solar Spectroscopic Imager: a purely scientific mission to understand the physics of particle acceleration and energy release in solar flares. The spacecraft carries instrument detectors which consist of high-purity germanium crystals. These are housed in cryostats which require to be kept at a very low temperature. The M77 cryocooler has performed that function to the satisfaction of the mission controllers for eighteen years at the time of writing and another has run for twenty-five years without degradation of performance.

In a less-exotic environment, Stirling machines are now in common use as efficient heat pumps. The principle is the same as for cryocoolers: an applied force will cause heat to flow from expansion space to compression, so that a Stirling engine with one end outside the house will pump heat to its other end inside. In this sort of application, Stirling pumps compare favourably with more traditional heat pumps. The one area in which Stirling seems unlikely to compete is in domestic refrigerators. The Rankine expansion-and-condensation cycle is simple and cheap to make and seems unlikely to yield to Stirling in the foreseeable future, so long as the gases it employs remain acceptable from an environmental point of view.

Overall, it is in the matter of cooling that the Stirling engine has had its greatest success, if measured by the simple number of appliances which use it. (There are a lot more coolers than there are submarines.) There is, to date, one other application to which William Beale's take on the Stirling engine seems to be uniquely suited. It is by far the most extravagant and improbable application, certainly as it would have been viewed by Dr Stirling. That is in NASA's KRUSTY generator, which is the subject of our last chapter.

CHAPTER 9

The Star Drive

It was my intention, in this final chapter, to begin by comparing NASA and its Voyager missions to previous explorations. Any comparison would have to be couched in terms of significance and size, but it is difficult to find an enterprise which qualifies on both grounds. The discovery by Europeans or Vikings of the Americas was surely significant – especially if you are American – but in terms of scale it didn't amount to much: Eric's few long-ships and Columbus's caravels were tiny compared with the significance of their ventures. Zheng He's seagoing junks meet the criterion as far as sheer size goes – it was a perfectly enormous fleet which he sent to Africa in the early Ming period – but his trip had few long-term consequences. (Shortly after it, the Chinese Empire turned its back on the rest of the world for half a millennium.) Admittedly, the size of his venture was paltry judged in the perspective from Pluto, but in relation to the society and economy from which it emanated, it was huge. It's the old problem of scale, in the same way that picturing the USS *Ronald Reagan* is: unless you have experienced it or something very like it, the imagination simply cannot cope.

You may have noticed in preceding chapters that *scale* is an issue which keeps arising. It arises with respect to the stars and, to a slightly lesser degree, the Solar System: evolution simply hasn't equipped our imagination or our perceptual apparatus to cope with the distances. We can use mathematics to compute them and pictures to make them seem familiar, but neither course amounts to a sympathetic understanding of what's involved. The human voyage into the space beyond our atmosphere is not just a venture into the unknown: it is, in a sense, a venture into the unknowable. If we survive the next few centuries, it will no doubt be viewed as the end of an era, as well as the beginning of another. We can be reasonably sure that, like Eric's or Columbus's voyage, it will be seen as astonishingly courageous – and, possibly, impertinent.

If this seems like overstating the case, we should consider what our understanding of the universe amounts to. Spaceships visiting stars and planets are the commonplace of film, television and comic books and are, thus far, familiar. But their familiarity comes with a cost: the condensation of apparent distances and timescales to a human level. Or the adoption of imaginary technologies whose only recommendation is dramatic convenience. Once you start to look at the matter in terms of the science, it becomes progressively harder to comprehend – for the science itself gets harder to understand: we have come a long way from the intuitive certainties of Newtonian space and mechanics.

So in addition to the problems of scale, we must add those of comprehension. But let's stick with scale: it's more amenable. Even there, though, you have to make an effort: take, for example, the hardware which has been developed to put people in space, and the institutions necessary to create and maintain that hardware. (This may seem to be taking us a long way from Dr Stirling's engine, but if we are to grasp the significance of what he did, it is necessary. Again, the problem of scale.)

The Russians' early ventures were bold, especially given the state of the Soviet society and economy at the time. The great enterprise of Communism had been launched on the back of Modernist optimism about the future: life for the masses was to be made better through both social and technological revolutions – and the latter were to be the handmaid of the former. When the standard of living of ordinary people failed to match that of the Capitalist West, the dream was kept alive by saying the millennium was only postponed and would be realised through superior Soviet technology. The space program was by far the most visible component of superior Soviet technology, especially since its ventures were amenable to display on television.

For more than two decades the Soviets would make most of the running in space flight. Apart from a few American ventures in the later 1940s, all the significant missions would be Russian. (They were not the first in space: that honour belongs to the Germans. Using a V2 rocket, in 1942, they sent a ship past the Karman line and so into space.) The Americans, though, would be first in the race to capture the Nazi rocketry and the early American flights were on modified V2s.

The Soviets were quick off the mark, however. With the help of the German technology, they had produced an ICBM by 1956 and the following year sent a dog called Laika into space. And that same year, Sputniks 1 and 2 would astonish the world. In 1959, the Soviets launched a rocket which reached a velocity sufficient to escape Earth's gravity. They then sent Russian flags to the Moon and photographed its far side – and, in contrast with their normal secrecy, released the photos to an astonished world. In 1961 would come the first human-crewed spaceflight. In that period, NASA had the consolation of taking the first photograph of the Earth as seen from orbit, but as the Cold War increased its icy grip, that would come to be thought inadequate by the American public and the US would enter the space race in earnest. From then on, space exploration would proxy for ideological warfare and the space race would parallel the arms race.

The size of the enterprise changed by orders of magnitude once it became the defining project of the United States of America. The National Aeronautics and Space Administration, established in 1958, must surely rank as by far the greatest technological venture in the entire history of the human race. In researching this book, I have been constantly astonished by the vast scale of its organisation. It is, ironically, like a set of Russian dolls, each of which opens to show another, smaller doll. Its budgets are such as many nation states would envy and it permeates American industries, universities and institutions like the roots of a mighty tree. It is an organisation on a scale which it is hard to relate to most people's lives. The list of crewed missions alone is indicative:

1959–68: The X-15

An experimental rocket-powered hypersonic research aircraft designed to be drop-launched from a Boeing B52 aircraft at 500 miles per hour at an altitude of 14 kilometres. The aircraft would fly at 4,500 mph and reach an altitude of 108 kilometres. It was the test bed for a number of control systems, of which control jets for vehicle orientation was the best publicised.

1958–63: Project Mercury

This had little to do with Mercury and a lot to do with being first to put a man into orbit. In this it was unsuccessful, for Yuri Gagarin became the first to orbit Earth in April 1961. This rather put America's fifteen-minute crewed space flight that year in the shade, but in 1962 John Glenn, aboard a craft propelled by an Atlas rocket, did three orbits, which saved a lot of American face. The state of the technology is apparent from the fact that the complex calculations of trajectory for this flight were done by three women rather than by computers.

1961–6: Project Gemini

This was a direct – and unsuccessful – attempt to take the lead from the Soviets by adding extra-vehicular activity, space rendezvous and docking to the support effort for the Apollo mission. Ten flights in all began in 1965, too late to upstage Gagarin's first flight and extra-vehicular activity in 1965 aboard a Vostok spacecraft.

1961–72: Apollo Program

This was the event which restored American self-esteem.

Public opinion was so incensed by the failure to match Soviet achievements that President Kennedy asked Congress to approve and fund a space research program on a scale never before imagined, and to commit to land a man on the Moon by the end of the 1960s. At a cost of over $20 billion (at 1960s valuation – multiply by ten to get an idea of what that means in today's money), it was the most expensive operation since the Manhattan Project. The program would use the huge Saturn rocket to put a two-part spacecraft into lunar orbit. A lunar module would land men on the Moon and a command module would return them to Earth. Four missions would be necessary to prepare for the lunar landing and a fifth would execute it, landing first Neil Armstrong and then Buzz Aldrin on the lunar surface. They would bring back rock samples and the project would contribute a vast fund of scientific data. Five further missions would also land men on the Moon. The whole thing was a huge success and it allowed Americans to congratulate themselves on achieving the first big prize in the space race.

1965–9: Skylab

This was an independent American space station. It weighed 77 tons and was put into a low orbit by a Saturn V rocket. It lost some of its insulation and a solar panel on launch but was repaired by its first crew. It was occupied by three crews for a total of 171 days, contained a laboratory for the study of microgravity and acted as a solar observatory. At 345 kilometres its orbit was too low for safety and NASA planned to use a space shuttle to take it to a higher, safer orbit. This could not be done in time and the Skylab re-entered the Earth's atmosphere in 1979. It was not accounted a great success, though a lot of useful information was gathered.

1972–5: Apollo-Soyuz Test Project

This was the result of a top-level agreement between the Americans and the Russians to fund a jointly crewed mission which would prepare the ground for what would become the International Space Station. Both nations were feeling the pinch by then, for space exploration was proving horribly expensive, especially to the Russians, whose economy was still failing to produce the promised affluence. It was agreed that in future the design of space-craft would be such that all American and Russian craft would be capable of docking with each other. In 1975 an Apollo command module met with and docked with a Soyuz craft.

1975: Viking

The Viking Project found a place in history when it became the first US mission to land a spacecraft safely on the surface of Mars and return images of the surface.

1972–2011: Space Shuttle Program

This was designed as a re-usable vehicle which could dock with and service the projected International Space Station. It would also function as a standalone research facility. It would be able to return to Earth at the end of a mission and, after being serviced and refuelled, make additional missions. The first one was launched in 1981, some nine years after the commencement of the program. It consisted of a spaceplane orbiter which carried the cargo and crew, attached to an external fuel tank and a solid-fuel launch rocket on either side. (The fuel tank is much larger than the orbiter – reflecting a major propellant issue which we will come to later.) The crew are from two to five persons, the maximum cargo load about 24 tons, and the attainable orbit between 185 and 643 kilometres, with missions lasting from five to seventeen days. The cargo bay carried Spacelab, a joint US–European Space Agency project, on twenty missions. In 1990 it launched the Hubble Space Telescope and in 1993 repaired it. In 1995 agreement was reached for the Shuttle to operate together with the Mir space station and in 1998 the former docked with the latter. There were two major disasters: Challenger in 1986 and Columbia in 2003. Nevertheless, over the period the Shuttle flew a total of 135 missions and was counted a great success.

1989: Galileo (Mission: Jupiter and Europa)
Galileo orbited Jupiter for almost eight years and made close passes by all its major moons.

1993 to the present: International Space Station
This was a gigantic effort involving NASA, the Russian Federal Space Agency (RKA), the Japan Aerospace Exploration Agency (JAXA), the European Space Agency (ESA) and the Canadian Space Agency (CSA). It is expected to run until 2030. It was the result of both the US and Russia wishing to share the burden of cost, and the desire on the part of the other partners to have a share in an enterprise which otherwise they would be quite unable

to afford. The US had originally projected a standalone Space Station Freedom but by 1993 had agreed to take part in a multinational effort. (There has to be suspicion, given the name, that the project was conceived more as propaganda than as a serious go-it-alone venture, for the budget was likely to exceed even the capability of the vast US economy.)

The ISS is an astonishing and untidy collection of modules accreted over the decades, every single part of which consumed resources of time, money and expertise which could never have been imagined in less-expansive times. Its components are constructed in factories and laboratories around the world, shipped into space and assembled. The assembly began in 1998 and is expected to be complete around 2030. It is serviced by Russian Proton and Soyuz rockets and by American Space Shuttles. The usual complement of the ISS is six or seven, but it can accommodate thirteen.

2017 to the present: Artemis Program

What would become the Artemis Program was partly the result of debates about the desirability of state-funded (as opposed to privately funded) research. NASA was the child of a society which was content to see the state undertake the management and much of the implementation of research – and hence shoulder the risk. Private capital could benefit from the expenditure of state funds through the subcontracting of subsidiary research and construction projects to private firms. But technologies became better-established, perceived risk declined and moves were made to secure a bigger chunk of the action for non-state capital. This coincided with neoliberal economic policies and led to a diminution of NASA's active role in future space exploration, in favour of private enterprise. The Artemis Program, whose object was to have a new space station orbiting the Moon as a stepping stone to a mission to land humans on Mars, reflected this change. Artemis was to be a joint venture with commerce and the European Space Agency to land a man and a woman on the south pole of the Moon by 2024, and thereafter to develop the Lunar Gateway preparatory to a Mars mission. (There seems to have been no objection in principle to partnership with the statist ESA.)

The launch vehicle is to be a new heavy-lift rocket based on the old Space Shuttle and called the Space Launch System. The initial aim was to lift 130 tons into low-Earth orbit and 45 tons to a higher orbit. Cost per launch was estimated in 2012 to be $500 million, later modified to $876 million and later still to $2 billion. Yearly budgets are in the region of $2.1 to $2.3 billion between 2019 and 2023. Total expenditure between 2011 and 2018 has been estimated at $14 billion. The infinitudes of space are not the only things that the ordinary human imagination has difficulty grasping.

But that is not all: the crewed voyages are only part of what NASA does. In the period since its inception, it has flown more than 1,000 uncrewed, robotic missions around the Solar System, visiting planets and asteroids and carrying out a vast range of scientific measurements and experiments. And in the course of the robotic, as well as the manned, missions, the science is being tested against observation and experience. It is an altogether unique conjunction: the science informs the engineering and the engineering, by its success or failure, provides verification (or otherwise) of theory. It is the greatest testing ground in the history of humanity.

All of which is very well, but we do seem to have departed from our purpose, which is to trace Dr Stirling's engine and show what became of it. Before we do, we ought to look at one aspect of space travel which affects distant unmanned missions.

—

There is a big problem of power for spacecraft travelling to the more distant parts of the Solar System. It is rarely the case that a craft can simply set its course and wait for its destination to approach. Course corrections are necessary, to cope with various forces which, if not addressed, will throw the ship off course. Corrections require energy and the energy requires fuel. The more course corrections, the greater the amount of fuel which must be carried. Photographs of the Space Shuttle on its launch pad show the small, elegant Shuttle vehicle stuck on the side of the big, ugly cylinder of the rocket which is to lift it off the ground. That huge cylinder isn't full of machinery: most of the space is taken up by fuel, all of which is to be used before the rocket is jettisoned. It follows that any fuel to be used to make course corrections will have to be carried inside the small spacecraft – which means there isn't going to be a lot of it.

And there is an arithmetical problem about carrying fuel for any purpose other than lift-off: the more additional fuel you carry, the more fuel you are going to need, to lift the fuel you are carrying into space. Which means you are going to have to carry still more fuel to compensate. And so on. There is a tendency for small changes in propellant requirement to run away and become exponential, so propellant budget is a big issue with all space travel. And the more distant the destination, the worse the problem becomes.

The propellant used for lift-off from Earth is chemical and the speed at which the burnt gases are expelled from the ship is determined by their chemistry and the design of the rocket. But there is a limit to how fast those gases can be expelled and this limit is not great. For all the fancy physics of the twentieth century, we still are limited by Newton's Second Law of Motion: force equals mass times acceleration. If a much greater acceleration can be achieved, a correspondingly smaller mass will do the trick. Since there is little prospect of achieving this with chemical propellants, we must look elsewhere. The only alternative presently in view is ion drive, in which tiny amounts of ionised gas are propelled at prodigious velocities. But an ion drive requires a supply of electricity. And this is where our Stirling engine comes in.

The matter of powering course corrections becomes more acute with distance: there are more opportunities for adverse influences to require corrections. And there is another, quite different consideration. Spacecraft on distant missions generally do not travel in straight lines, for they must make use of a tactic variously called swing-by, gravity assist and slingshot. This device goes back a long way and, like so much in space technology, has Russian ancestry. About 1918 one Yuri Kondratyuk suggested that it might be possible for a spaceship to hitch a lift on a passing planet by being carried along by the planet's gravity. This is, indeed, possible, and was first used by the Russians in 1959 to carry Luna 3 to the far side of the moon.

All the planets in their orbits travel with a velocity relative to the Sun. If a spacecraft can come within the gravitational field of a planet, it can be carried along with it and have its speed increased in the direction of the planet's travel. Obviously, it will need to expend energy to break free of the planet's gravity but the energy required to do so will be less than the energy gained. The same device can be used to change course or even to decelerate.

The downside to this scheme is that planets are rarely in the right place, travelling in the right direction for this to be of any use. But in 1964 an engineer called Gary Flandro, at the Jet Propulsion Laboratory, worked out that in the late 1970s the outer planets would be in a position which would allow a craft to visit them all in a single trip. This caused much excitement at NASA, as you can imagine: not least because Mr Flandro's calculations also showed that the alignment would not happen again for about 175 years. It was an opportunity not to be lost and NASA prepared for the greatest voyage yet. It was to be called the Grand Tour Program.

NASA created an outfit called the Outer Planets Working Group, which, after deliberation, proposed two missions, each of which would visit three outer planets. The first would go to Jupiter, Saturn and Pluto and the second to Jupiter, Uranus and Neptune. The proposal, undeterred by the cost, called for a new ship which was to have the snappy title of the Thermoelectric Outer Planets Spacecraft, or TOPS for short. (The 'thermoelectric' designation suggests that the source of their power out in the far reaches of the System would be plutonium.) By 1971 the projected cost had passed $1 billion, which proved too much for the money men and the whole thing was cancelled.

In 1972 a more modest proposal met with approval. It was to be called the Mariner Jupiter-Saturn Mission and was to cost $360 million. Reading about the preparations, it is hard to avoid the impression that the project managers were less than candid with their superiors. The ship was to be built in such a way that it might well last long enough for the Grand Tour but, to reduce the estimated costs, was advertised as involving missions only to Jupiter and Saturn. Thus was born – or at least conceived – the Voyager Program.

There were to be two Voyager craft. Voyager 2 was, paradoxically, to leave first – but Voyager 1 would be first to attain the ultimate objective of leaving the Solar System. It would fly by Jupiter, gain a trajectory boost from Saturn, and another from Uranus, which would take it to Neptune and beyond. Voyager 1 would fly by Jupiter and then Saturn before passing Pluto on its way to the edge of the Solar System.

The whole thing was a great success. The gravity-assist accelerations and course changes worked perfectly and the Voyagers would become the second and third human artifacts to leave the heliosphere. (Pioneer had been first but was overtaken.) Both Voyagers carried messages of goodwill for any extraterrestrials they might meet in their near-eternal passage in the void.

—

In May of 2018 NASA announced the successful test run of their KRUSTY generator. KRUSTY (Kilopower Reactor Using Stirling Technology) is an electricity generator which consists of a dustbin-sized nuclear fission reactor, atop which sit eight free-piston Stirling engines, each equipped with a linear alternator. The fission reactions provide the heat to run the Stirling engines, each engine drives its alternator and the alternator produces current. The whole thing is small enough to be loaded on a rocket and sent to the Moon and other, more distant, places.

KRUSTY was designed to meet a felt need. From the beginning of manned space flight, it was apparent that there was a requirement for on-board electrical generating capacity – and there were few obvious sources from which this might be obtained. When you consider it, people on Earth are dependent on a fairly small number of sources for their electricity. We have power stations – coal, oil, gas, nuclear-fired – which are pretty similar in the way they work: they all generate heat, which by making steam is turned into mechanical motion and the mechanical motion makes electricity by spinning coils of wire and magnets past each other. Wind generators and hydro-electric stations do the same, only using wind or water instead of heat. Geothermal generators make steam in their bowels. We have photovoltaic generators which turn sunlight directly into electricity. A few tidal generators. That's about it, apart from very minor sources. All of them, apart from the photovoltaic (and until May 2018 the nuclear), are quite unsuited to generating power in space.

In the last thirty years or so, artists' impressions of settlements on the Moon and Mars either don't indicate any power source or show lots of solar panels. Photovoltaic panels work pretty well on the Moon, though you would need a lot of them to provide enough electricity for a bunch of astronauts to live in any comfort. They would also have to provide battery power for all the things the astronauts are on the Moon to do. And of course you have to get them there, which isn't easy since solar panels are heavy and bulky. But the big problem with lunar photovoltaics is the inescapable fact that for much of the time it's night on the Moon and solar panels need sunlight. You can charge batteries using the photovoltaic sources, but batteries are very heavy – and you have to get them there, too. As regards distant space travel, there is another, bigger problem for photovoltaics: the strength of the Sun's radiation diminishes with distance from the Sun. Worse: it diminishes as the square of the distance, so by the time you get out to the orbit of Jupiter, your solar panels aren't going to work at all.

Toward the end of the twentieth century, NASA had begun to address this problem – for on even the worst scenario, the time was going to come pretty soon when they would have to think about making a settlement on the Moon and possibly Mars. (That's possibly unfair to the folk at NASA: they had always had these long-term objectives. It is, however, true of the politicians who provide the money which NASA needs to realise its ambitious aims. We should root for the Russians, Chinese et al: without the compulsion to outdistance their competition, the cash for space exploration in the US would probably have dried up long ago. National egotism has its uses.)

One possible source of space power presented itself fairly early. Some elements on the far side of uranium, and isotopes of those elements, generate heat spontaneously as they decay. (They also generate other, less-desirable kinds of radiation, but that's beside the point, though not entirely irrelevant.) It was decided that the most promising of those elements was an isotope of plutonium, Pu238. As it decays, Pu238 emits alpha particles which, when they collide with their surroundings, produce heat. Since the half-life of Pu238 is 87.7 years, it furnishes a long-lasting and conveniently compact source of intense heat. (It is also phenomenally toxic, chemically and radiologically: while the alpha particles very soon decay, if the stuff is inhaled or ingested the result is about twenty times worse than an equivalent dose of gamma radiation. And in terms of everyday chemical toxicity, plutonium makes cyanide look like mother's milk. Not something you want to share a confined space with.) But when Pu238 is situated adjacent to a thermocouple (a strip of two different metals fused together) the alpha radiation will decay to heat and the heat will cause the thermocouple to produce an electric current.

Because Pu238 is such nasty stuff, NASA long ago parcelled it up into a form which would resist accidental damage and, at the worst, a spacecraft crashing into Earth and releasing its burden of plutonium. The result was the General Purpose Heat Source, or GPHS: a strong (very strong) box in which are Pu238 pellets which have been fabricated into ceramic pellets of plutonium dioxide and are encapsulated in iridium sheaths. The whole thing is designed to survive catastrophic involuntary re-entry.

When a GPHS is put together with a thermocouple, it becomes in NASA-speak a Radioisotope Power System, or RPS. (NASA use lots of acronyms. You can't blame them, given that our ordinary language struggles with some of their designations.) RPSs have been the power source for most of the successful missions over the last few decades: Apollo, Viking, Voyager, Galileo and many others. They are very safe, very reliable and they have long lives. The Voyager missions depended on RPSs (remember the TOPS, the Thermoelectric Outer Planets Spacecraft) and continued to transmit data long after they had left the Solar System. The big problem with RPSs is that they don't produce a lot of electricity and craft on distant missions must function on power levels which the average domestic kitchen would find inadequate. The trouble lies mainly with the thermocouple, which isn't an efficient convertor of heat to electricity.

In the 1970s NASA's Glenn Research Center (then called Lewis) decided to look at using Stirling generators instead of thermocouples. (The output of a Stirling generator even then was about four times that of a thermocouple.) NASA had had a connection with Sunpower from near its inception: the latter were very small and very unconventional compared to the corporations which were NASA's main suppliers, but they were producing some new and potentially useful ideas, and NASA were interested in their Advanced Stirling Convertor (ASC). The development wasn't quick (remember the quip about the simplicity of the free-piston Stirling engine being illusory), but in October 2007 Sunpower delivered to NASA two ASC-E convertors – right on schedule. (They also sent one to the Department of Energy as a spare, just in case.) The converters were the product of a long and difficult process of development and testing, but by the time they were delivered they could be considered a mature technology. This meant that they could be guaranteed to perform at least as well as the contract specified for a very long time without repair or maintenance. It had been anticipated that the ASC would deliver approximately twice the power of a thermocouple generator, at seven watts per kilogramme. They delivered eight. Sunpower would continue to develop the converter, through another four generations, each an improvement on the previous. As usual with free-piston Stirling engines it isn't much to look at, but it was the product of forty-three years of intensive effort.

In the early '90s, using the ASC, NASA began an in-house programme to develop a system which would become the Advanced Stirling Radioisotope Generator, or ASRG: they were confident that Sunpower would deliver the generator part. The ASRG is a 55-watt converter powered by a General Purpose Heat Source. Besides producing more power for a given weight, the ASRG was more economical of the Pu238 fuel than the thermocouple-based RPS, though the latter was to continue in use for some years for small applications.

From 1964 until 1988, the US would produce all the Pu238 it needed at the Savannah River Site weapons-grade reactors. When production ceased at Savannah River in 1988, the Department of Energy held a stockpile likely to suffice for any foreseeable needs. But with NASA planning ever-more ambitious missions, it soon became apparent that another source of supply would have to be found. The only possible source was Russia, but by that time the Soviet Union was in chaos and there was doubt as to whether central government would be re-established. It *was*, but would be central only to Russia and it would be some time before a central authority would be able to strike a deal to supply Pu238. (It would be interesting to have the details of this and of its implementation, given kleptocratic control of the Russian economy at the time. Particulars are hard to come by and there is suspicion in some quarters that large amounts of currency were involved. Certainly, there are reports of illicitly obtained quantities of Pu238 being discovered by the authorities.) In 1990 an agreement was made whereby the US Department of Energy would buy from Russia up to 40 kilogrammes of Pu238 over a period of five years at a cost of $57.3 million. From 1993 onward, all the Pu238 used in American spacecraft came from Russia. But with the resurrection of the Russian state and the renewal of its ambitions to engage in space exploration, the supply was eventually cancelled. (It would be blamed on American failure to create facilities for destruction of weapons-grade plutonium, which was true enough, but was probably the result of a general deterioration of relations between Russia and the US.)

This caused concern for the ASRG project, and it would be cancelled in 2013, because the planning of missions was becoming conditional on the availability of Pu238, which was worse than doubtful. By 2015, the total US inventory of Pu238 available for civil purposes was 35 kilogammes, of which only about one kilo was in good enough condition to meet NASA's specifications. There was some noise about this in the press, but surprisingly little, considering how much national prestige was involved. The US eventually recommenced production – in an uncertain and disconnected sort of way – and are now reported to be making about 400 grams per annum. Not a lot, considering that each GPHS uses about a quarter of a kilogramme.

Though nobody said so at the time, the KRUSTY project was almost certainly related to the prospective shortage of Pu238: Uranium suitable for use in fission reactors was not in short supply and if fission could safely replace radioactive decay as a power source, a lot of worries would be laid to rest. Reports commissioned by NASA in 1989 and 1990 had recommended Stirling generators as feasible for the production of power for space use. In 2003, NASA had funded Sunpower to develop the Advanced Stirling Convertor for that very purpose. (Though the generator was to be used first in the ASRG.)

In 2010 NASA had commenced what they called the Planetary Science Decadal Survey. This was a study to assess whether fission reactors might provide an alternative to radioisotope-based power systems. By 2012 they reported success in a test using a Flattop Critical Assembly. The latter is a container specially designed to contain fissile materials which are allowed to go critical to the point of creating heat, but not permitted to turn into bombs. This study was quick – and cheap enough for the report to mention the fact: it took only six months and cost less than a million dollars, small change by NASA standards. The test reactor produced 24 watts, about enough to run a bicycle lamp: probably the most expensive watts in history. The demonstrator had a highly enriched uranium core with a central hole to accommodate a heat pipe which conveyed the heat to two Stirling convertors. The heat pipe was a technology which Sunpower had developed and this was the first time it had been used in a fission reactor. It was also the first time a Stirling convertor had been run off a nuclear fission heat source. Dr Stirling would have been pleased.

In 2014 the Nuclear Power Assessment Study was commenced: its remit, to determine whether a small fission reactor would be possible for planned missions such as the Titan Saturn System Mission and the Uranus Orbiter Probe. The same year the Evolvable Mars Campaign began: a truly astonishing enterprise which is the commencement of an eventual colonisation of the surface of the red planet. And, in NASA speak, 'small fission power baselined for pre-crew propellant production and post-landing crew operation'. (I'm fairly sure that means they are going to use a KRUSTY for power to get about on Mars. What the pre-crew propellant is, is anybody's guess.)

By 2015 the free-piston Stirling engined generator could be considered a mature technology, in that all the main problems had been solved and such a generator had been running continuously for some years without problems or maintenance. That same year NASA commenced the Kilopower Project. It was given three years and up to $20 million to design, build and test a prototype reactor. The participants in the project were (of course) Sunpower Inc., of Athens, Ohio; The DOE Nevada National Security Site, where the nuclear materials were kept; the DOE Los Alamos National Laboratory, where the reactor assembly and testing was done; NASA JSC, the L. B. Johnson Space Center, where mission flight control is located; NASA MSFC, the Marshall Space Flight Center, which does research in rocketry and spacecraft propulsion; NASA Glenn Research Center; Advanced Cooling Technologies, Inc.; and the DOE National Security Site, which is the birthplace of the atomic bomb and has a facility for enriching uranium and suchlike innocent pursuits. A truly colossal enterprise, which can be forgiven a few acronyms.

And KRUSTY, what does it look like? From a distance it looks like an umbrella or a parasol. It has a stem, at one end of which is a set of vanes which, folded, give the umbrella impression and are raised when the thing is deployed. At the other end is a core made of a uranium molybdenum casting, which is the bit which generates the heat. It is surrounded by a neutron reflector of beryllium oxide, with lithium hydride/tungsten shielding above. Rising out of the core are eight heat pipes containing liquid sodium, each of which leads to a Stirling power convertor. The waste heat is conveyed to the umbrella vanes which, when unfurled, act as a radiator, dispersing the surplus heat.

The reactor is intended in the first place to produce about one kilowatt of electrical power, though the makers expect the design to be capable of enlargement to 10 kW without undue scaling problems. It will produce at least ten times more power than a thermoelectric generator. It is intended to facilitate operations on the surface of the Moon and, eventually, Mars. Other probable locations are Europa, Titan, Enceladus, Neptune, Pluto and others. Should anyone be daft enough seriously to contemplate mining operations on asteroids (some indeed do) then no doubt it will come in handy there.

It now seems fairly clear: all the developments envisaged as regards distant space flight and extra-planetary settlement will require a source of onboard power and, as things stand at present – and as they look likely to stand for any foreseeable future – that power will be provided by a Stirling engine. But we are lacking in two factors which must be addressed if the engine is to qualify as a star drive. Firstly, it must be used for propulsion – and so far, all the applications of the technology mentioned have been ancillary to the matter of propulsion. Secondly, the stars. Dr Stirling was a keen astronomer and no doubt would have resisted any confusion between stars and planets. Planets are the things which orbit our sun; stars are other suns and much, much farther away. Being the progenitor of a drive which will get us to the other planets would be regarded by most folk as a sufficient accolade; but could it seriously take us to the stars?

—

We now verge on a field in which fantasy is fuelled by wishful thinking and the inconvenient laws of nature are often set aside in the interest of plot development for novel, film and comic book. It's called science fiction and not to be sneered at, for it has produced some very important original ideas, as well as given pleasure to millions. But Dr Stirling's engine, however improbable and however ill-understood, is the child of science, not of fiction. So in what follows we will avoid unfounded speculation and stick as closely as we can to facts and testable theories.

As far as is known – and by 'known' I rule out mere speculation – Newton's Third Law of Motion is nowhere suspended. In the universe of which we have knowledge, it is always true that 'For every action there is an equal and opposite reaction.' Reaction is what drives rockets in space: if you are floating in space and you wish to move in a particular direction, you must eject a mass in the opposite direction. Your motion will then be a product of the mass ejected and the velocity (relative to you) of the ejection. There are no exceptions, none at all. Space rockets move by throwing burnt gases out of their tails: the more gases and the faster they are ejected, the quicker the rocket will move. If they want to change direction, they must do the same, but throw stuff out at an angle to their direction of movement.

The rockets, which we see on television lifting satellites and crews into space, work on the same principle, with the addition that, in the Earth's atmosphere, they have something for their propellants to push against. They must accelerate quickly to escape Earth's gravity, which is the reason why they make such a fuss. But once clear of terrestrial gravity, things can be more leisurely and alterations of velocity and direction made more slowly. A relatively gentle push for a long time will, in the absence of any countervailing forces, effect the same result as a big push for a short time. But Newton's Third Law is not violated: the velocity of the ship will depend on how much stuff it has to eject, and the speed at which it can contrive to eject it. Up until the present, the rockets lifting off the Earth's surface have been powered by chemical reactions which produce high-velocity gases. Despite NASA's best efforts, they are crude devices, close cousins to the fireworks with which we used to play as children. Some degree of control of the rate of burning their propellant is possible, but it could not be said to be done with finesse, and the whole thing is over in a matter of minutes.

Away from the excitement of rocketry, consideration of the stars is sobering. The sheer scale of the universe is conceivable only by the use of mathematical notation; not by imagination – for as terrestrial beings our ability to imagine distance has been determined by the size of the planet on which we evolved. Our ability to come to terms with the requirements for any significant travel beyond the confines of the Solar System seems to be similarly constrained, with the result that suggestions for how interstellar travel might be done mostly involve fantastic ways round the impediment imposed by natural law.

There is one type of star drive, though, which is not only possible but well-tried. That is ion propulsion. This has been known about since the late 1950s and was first tested in space by Glenn in 1964, under the becoming title of the Space Electric Rocket Test. The test was favourable and the technology is now routinely employed to keep communications satellites in position and even for propulsion, especially for missions which would be unable to carry sufficient chemical propellant. Ion thrusters have what in NASA speak is called 'high specific impulses', that is, they throw stuff out very fast, so they get the same push as they would if they threw more stuff out, but slower. The stuff in question is mostly xenon, a gas which is heavy and easily ionized.

The process of ionisation consists of bombarding the xenon with high-energy electrons so that electrons are knocked out of the xenon atom and it becomes a positively charged ion. The positively charged xenon atoms, together with the high-velocity electrons, become a plasma which can be manipulated by electric and magnetic fields. Positively charged ions are accelerated through a positively charged grid at very high voltage toward a negatively charged electrode. This allows them to attain very high velocities and become an ion beam. There is an arrangement whereby all the surplus electrons do likewise to produce an overall neutrality and a net thrust in the direction of the beam. The propulsion system was used for a Glenn mission called Deep Space 1 which, between 1998 and 2001, travelled more than 163 million miles and flew past the Braille asteroid and the comet Borelly. The Dawn spacecraft which launched in 2007 is propelled by NSTAR ion thrusters. It is the first craft to visit the asteroid belt between Mars and Jupiter.

NASA has developed an ion propulsion system which produces three times the thrust of NSTAR and has now run for nine or ten years continuously, demonstrating its suitability for distant missions beyond the confines of the Solar System. Other ion thrusters are currently in development which will offer thrusts in multiples of those obtained so far. There seems little doubt that future distant missions will be propelled by ion thrusters. And, of course, the Voyagers owed their original title to their possession of ion thrusters powered by thermoelectric generators.

Before we leave this topic, it may be a good idea to say a few words about possible alternatives to ion drive for deep space travel. After all, we are all familiar with *Star Trek*'s Warp Drive, a technology (if that's the right word) which permits faster-than-lightspeed travel. Why not consider Warp Drive? Why stick with boring old Newton when we can contemplate interstellar tourism? And anyone who reads popular magazines and science-fiction books will know that there are lots of possible ways around dull relativity. Indeed, doesn't Einstein say that space is curved: if it is, why can't we take short cuts between the curved bits and save ourselves a lot of time? (The reason, of course, is that if it's *space* that is curved, there isn't *anything* between the curved bits. Wishful thinking colludes with ignorance of basic cosmology to lead to a category mistake.) The same goes for wormholes.

Yet it seems that some people who should know better think the idea of faster-than-lightspeed travel isn't daft. In April 2010 the US Defense Intelligence Agency released a thirty-four-page report which they had commissioned, titled 'Warp Drive, Dark Energy, and the Manipulation of Extra Dimensions'. If such an important organisation thinks those sort of things are worth spending taxpayers' money on, surely there must be something in it? And the report thinks we may soon be able to manipulate dark energy to access higher, unseen dimensions which will allow us to move at speeds greater than that of light.

The report says:

> Control of this higher dimensional space may be a source of technological control over the dark energy density and could ultimately play a role in the development of exotic propulsion technologies: specifically, a warp drive ... Trips to the planets within our own solar system would take hours rather than years, and journeys to local star systems would be measured in weeks rather than hundreds of thousands of years.

That all looks promising – or it was, until it came to the notice of Sean Carroll, a theoretical physicist at CalTech, and very grounded despite his specialisms. 'It's bits and pieces of theoretical physics,' he said, 'dressed up as if it has something to do with potentially real-world applications, which it doesn't ... this is not something that's going to connect with engineering anytime soon, probably anytime ever.'[1]

Professor Carroll doesn't pull punches: he goes on, referring to the plethora of crackpot ideas which he calls 'a whole zoo of nomenclature devoted to categorizing all of the non-existent technologies of this general ilk ... there is no reason whatsoever why these claims should

---

1 Sean Carroll, 'Warp Drives and Scientific Reasoning', *Business Insider*, 26 May 2015.

be given the slightest bit of credence, even by non-experts
. . . for a very scientific reason: life is too short.'[2]

That about covers the field: exotic proto-technologies
are just that. Every one of the existing technologies has
a respectable origin in good science which itself is part
of the ongoing tradition of observation, classification,
hypothesis, experiment and critical assessment. The
problem for devotees of science fiction is that all of these
require old-fashioned hard work, even where they refer to
the most apparently exotic of disciplines.

The trouble with exotic disciplines is that they generally look exotic only to outsiders. I remember forty years
ago getting on a plane to London and finding myself
seated beside a very old friend called Kenneth, who was
on his way to CERN in Geneva. The conversation went
roughly as follows.

Me: 'Hello, Ken. Haven't seen you for ages. What are
you up to these days?'

Ken: 'Oh, same old stuff. Still at CERN.'

Me: 'What old stuff is that, then?'

Ken: 'Same stuff I've been working on for years: fun-
damental particles, what the universe is made of. Nothing
very exciting.'

What's exotic depends on who is looking. The same
is true, I suspect, of warp drives. The point about ion
drives is that they work. They sound exotic (if you aren't
used to hearing about ion drives for spaceships) but,
boringly, they do work: they push spaceships along and
have been doing so for years without anyone getting too

2 Sean Carroll, 'Warp Drives and Scientific Reasoning',
Business Insider, 26 May 2015.

excited about it. And as things stand, they look likely to propel our spaceships out beyond the solar system and maybe, eventually, to the stars. It won't be in our lifetime or that of our grandchildren, but it's exciting for all that. At least that's how it looks from where I'm standing.

And Robert Stirling's engine? Ion thrusters need electricity to make them go. There can be no reasonable doubt that, for the foreseeable future, ships sent into interstellar space from this planet will depend on propulsion by ion thrusters which themselves are driven by KRUSTY generators. Whether they will get to the stars is anybody's guess. But if they do, the first human mission to reach one will be powered by Robert Stirling's engine. Without the least exaggeration, a star drive.

Epilogue

Robert Stirling would continue his ministry in Kilmarnock for eight years. In 1823 Dr George Smith, the minister of the neighbouring parish of Galston, died and Robert was called to serve in his place. Since Smith had been a moderate in Kirk politics, Robert need not fear the turbulence which was customary in Kilmarnock. He did, though, have to cope with inherited fame again, for this predecessor too had been the subject of one of Burns' verses: this time from a famous poem, 'The Holy Fair'.

> Smith opens out his cauld harangues,
> On practice and on morals,
> An' aff the godly pour in thrangs
> To gie the jars an' barrels
> A lift that day

Like the best of Burns' poems, it's written in Scots. Its import is that Smith was a moderate who preferred morality to hellfire as a topic for sermons – and that the parishioners, who were very fond of hellfire on Sunday mornings, deserted the Kirk in search of spiritual comfort of another kind. It seems that Smith was offended, and said so, whereupon Burns replied with another verse, which alas is incomprehensible to those unversed in demotic Scots.

Galston was a large village about 5 miles up the Irvine valley from Kilmarnock, so the removal caused no fracture in Robert's relationships. By then he had become close friends with Thomas Morton and continued for some years to use the workshop which Morton had built for him in Morton Place. In Chapter 1 we mentioned the likely influence of John Leslie on Stirling's development: from his taking up his ministry in Kilmarnock, it is probable that Morton became a collaborator too. The move to Galston would prove an impediment to Robert's astronomy and his engineering works – but not much (though public transport was rudimentary, in those days 5 miles was not thought far to walk) – and in time he would have a workshop built close to the Manse in Galston, where he would work in the evenings. He is said to have built an engine – a Stirling engine, of course – which powered his lathe. In the early years of his ministry, his parishioners were said to have been disquieted by the nightly noises which emanated from the vicinity of the Manse. No doubt they got used to them, for Robert would work on engines for the rest of his life.

The village – it was a small town, really, with a population of about 3,500 – to which Robert moved in 1824 cannot have proved much easier for a new minister than did Kilmarnock. The people were occupied mainly as weavers and colliers. The weavers were hand-loom weavers: they lived, with their families, in two-roomed cottages, one room of which was taken up by the loom.

They were self-employed, weaving cloth on commission for itinerant contractors. Until recently, demand for their services had exceeded supply and they had been prosperous. The combination of (relative) prosperity with a long tradition of independence and ideas of equality and liberty had caused them to be boisterously attached to political radicalism. They were also remarkably well-informed: literacy was another of the old Scots traditions and many of them owned small libraries of books. There were, in addition, lending libraries from which they might borrow works on current affairs. There was a strong current of anti-clericalism which would pose an obstacle to any new, young minister.

The advent of power looms in nearby Kilmarnock was rightly seen by the weavers as a threat to their livelihood. The looms in Kilmarnock in 1824 were mainly water-powered, but before long steam engines would be introduced and factory production become the norm. Hand-loom weavers were unable to compete with steam-powered factories and their living standards would decline. Along with that decline would go a diminution in religious observance. They fiercely resented their relative impoverishment, as they did the necessity to take employment in the factories. Factory labour in the early nineteenth century was the antithesis of the freedom which the self-employed weaver so cherished. Hours were long; hard, monotonous labour continuous and docility a condition of employment.

The geology of Ayrshire is such that coal seams were to be found close to the surface in many places and small drift mines had been dug. As demand for coal increased

214

with the advent of industrialisation, shafts were sunk in many places around Galston and colliers recruited from among the unemployed. Colliers were regarded by weavers as belonging to a lower social class – which indeed they were, judged by almost every criterion, having been recruited mainly from among the indigent population. With less social organisation and education than the weavers, the colliers were not so much anti-clerical as simply irreligious.

—

We are fortunate in having a detailed description of Galston parish, written by no less than Robert Stirling himself. It is part of the *Second Statistical Account of Scotland*. The *Statistical Account* had been a project of Sir John Sinclair's toward the end of the previous century. Sir John was a very busy landowner and lawyer from Caithness in the far north of Scotland who moved in governmental circles in Edinburgh and London. An enthusiastic proponent of agricultural improvement, Sir John believed that the rational organisation of human affairs was a prerequisite of progress. To provide a baseline for estimating the effect of interventions, he proposed a statistical survey of the condition of the country. (He was the first to introduce the term 'statistical' to the English language, though the meaning he attached to it was a little different from ours: more an estimation of happiness.) He devised an original scheme of asking all the ministers in the country to compile a statement of their parishes along lines which he would lay down. The *First Statistical Account* would be dated 1790 – and much of the information we have about Robert Stirling's early circumstances is culled from the *Account of Methven* parish. A second *Account* would be drawn up in 1843. Robert Stirling would contribute the description of his parish of Galston. So the following are some extracts of his account of his parish, by his own hand.

He begins by deploring the imperfect state of the Parochial Registers: prior to the year 1592, they are in a mutilated condition, though not so bad that he is unable to tell us a good deal about the Roman occupation some 1,600 years before. He also relates the natural and civil history of the parish, before turning to the state of the population. He is much concerned with the weavers, some of whom he says are turning to the weaving of fine silk. (This was a small market which machine looms were temporarily unable to serve.) 'The high wages,' he says, 'which could formerly be earned by weaving or sewing have introduced among this class a taste for an expensive mode of living, which contributes greatly to abridge the real comforts of life, when wages are verging, as at present, towards the lowest ebb. Their condition, therefore, may now be reported as far from comfortable, and the discontent naturally arising from this state of things has been greatly increased by the ignorant or dishonest labours of political agitators, who have taught them to ascribe to oppression and misgovernment, what is chiefly owing to the multiplication of power looms, and other machinery.'

Stirling would be concerned for the rest of his life with the condition of his parishioners. Being their minister, his concern would be firstly with their spiritual well-being

– but he plainly believed there was little point in preaching redemption to the starving and therefore spent much, perhaps most, of his ministry in tending to their secular welfare. For this, we have the word of one of his successors in 1909 who, modestly, remains anonymous:

> The close of the year 1848 and the beginning of 1849 brought a terrible calamity to Galston and the country. The scourge of cholera came. It was then that Dr Stirling showed the stuff of which he was made. Then did he manifest the fidelity and Christian courage of a true minister of Christ. Fearlessly he moved among the plague-ridden homes of the parish; faithfully did he minster to the physical and spiritual wants of the sufferers. He toiled among them night and day; he tended them; he prayed with them; he buried them. And I am only speaking the literal truth when I say that his conduct at that time won something nobler than the presentation which was given to him for his faithful and devoted service, and that was the lasting gratitude of many a soul in this parish. When I think of these days and when I think of that work so faithfully and fearlessly performed, I have no hesitation in characterising this country minister as a Christian hero.[1]

You may have noted that the reference is to Dr Stirling. Around the year 1840, Robert Stirling was awarded

---

1 *The History of Galston Parish Church*, 1909.

the degree of Doctor of Divinity by the University of St Andrews. This we know, but unhappily little else. The DD was an honorary degree and there can be no doubt that it was awarded, but St Andrews University seems to have no record of it – or if they do, none that I have been able to discover. We know that Robert was a considerable biblical scholar: extant sermons attest to that. But he does not appear to have written a great deal on biblical matters: again, there may be texts, but none discoverable by me. The consensus of opinion is that the award of the degree was not for any specific action but simply in recognition of his exemplary services as a parish minister. *The History of Galston Parish Church* attests to that.

And note: they gave him the degree *before* the main cholera outbreak, so we may presume that his care for the cholera victims was not an isolated example of his goodness. First reports of cholera in the Far East had arrived about 1830. Its cause was then unknown, but it travelled fast. True to form, it arrived in Glasgow in 1831 and caused death, panic and despair in the overcrowded slums. The people of Galston managed to escape the worst of the earlier outbreak, but in 1848 they succumbed on a large scale, with huge mortality. Cholera is a bacterial disease, spread mainly in drinking water, but the bacillus can survive wherever there is moisture – and in damp, dirty, overcrowded housing it spreads rapidly. It causes diarrhoea, which, in the absence of adequate drainage and sanitation, enhances the spread. The effect in Galston was deadly. Nobody knew why the disease had come or how it was transmitted: all they could do was tend the sick until they died – which was a disgusting as well as a dangerous thing to do. Merely to be present was to be exposed to the pathogen. Little wonder that the folk of Galston came to view their minister in more than favourable terms.

Not everyone did, though. In 1834 there began a perfectly fantastical squabble which was to involve the General Assembly of the Kirk, the Court of Session in Edinburgh, the House of Lords and, eventually, Robert Stirling. The parish of Auchterarder had been without a minister and the patron of the parish, the Earl of Kinnoull, exercised his undisputed right of patronage to present one Robert Young as minister. Young was a well-known drunkard and the only parishioner to sign the petition for his installation was the village publican. The great majority of the people objected and a dispute ensued which would bring into question the nature of the relationship between the Kirk and the secular state. I would summarise what then happened if I could, but the complexity of the arguments over the next seven years is such that I confess myself unable to do so: I think that today, only someone steeped in the intricacies of theological minutiae is likely to be able to oblige. Robert evidently had no such problem. He considered himself obliged to minister to the clergy in the parish of Strathbogie, who had been forbidden their churches by the General Assembly of the Kirk. This would lead to his being suspended by the subsequent General Assembly, the governing body of the Kirk, from participation in the work of his presbytery.

The Great Strathbogie Controversy, as it came to be known, was about high matters of principle, of a sort that few people today would relate to. Though Robert was severely censured, and did not resist the censure, absolutely nobody seemed to hold it against him: both sides of the argument were agreed on that. So, though for a short while he was officially an outcast, his reputation in the Kirk as a whole did not suffer. In his parish, it goes without saying, he was a hero.

Robert Stirling would stay in Galston for the rest of his long life. He would have a happy marriage and seven children, of whom three boys would become railway engineers. He would continue to work on his engines in the workshop beside the Manse and, as we have seen, would enjoy the esteem of some of the great engineers and physicists of the nineteenth century. Otherwise, he continued to do what he had long ago decided to do: be a humble minister in a small-town parish.

He died one of the best-loved ministers in the Church of Scotland. A photograph, taken in old age, hangs in Galston Parish Kirk, where Robert preached. Nearby there is a small memorial in the form of a model Stirling engine. Members of the Kirk of Scotland never went in for pilgrimages, regarding them as products of prelatic superstition; there is, nevertheless, a small but steady stream of visitors to Galston kirk. As the author J. R. Senft writes, the engine which Stirling designed was 'in advance of practically every single design produced during the following 100 years'.

Stirling was, by any reasonable standard, a good guy. If he hadn't been a Presbyterian, they would probably have made him a saint. His engine was remarkable; the economiser was inspired. We have seen that he didn't know why it worked, but, on some level of understanding, he could see the need for it, and devised the instrument to serve that need. Perhaps the question we should ask is a different one: about our understanding of what we mean when we say that we understand something. Is there a kind of comprehension which runs deeper than that which can be explained in perfectly explicit terms? If there is, then Robert had it early on, when he designed the 1816 engine.

The parallels with William Beale are strong: both were quiet men who pursued their vocation; neither sought fame nor wealth; both had a strong sense of what is right. And both saw in their engine a way of benefitting their fellows. I think neither would be disappointed in it. I had thought that a suitable ending for this book would be an imaginary meeting in some celestial workshop between Beale and Stirling, in which they discussed how NASA should have gone about employing their engine. It was tempting, but would have been mawkish – which would have done justice to neither. Pleasing to think about, though: the shades of two old engineers watching their offspring as it orbits Alpha Centauri.

—

So that's it: the story of Dr Stirling's engine. It's a far cry from Galston to the stars, but with any luck the engine will get us there. That seemed a pretty neat ending for a book: success doesn't come much bigger than that, measure it how you will. And surely it's dramatic enough, given the long list of disappointments, to end on a high like that. But if we are to be candid, we must admit that there are still a few things hanging in the air. What is likely to come next? Let's hope not another disappointment.

Happily, there seems to be no likelihood of that. The development of the free-piston engine didn't just stop when Sunpower and NASA got it to do what they wanted it to do. In their hands, and in the hands of others, the improvement of the free-piston Stirling engine is continuing and there is no telling what form it may take. Nor did the improvement of the kinematic Stirling cease when Kockums installed it in their submarines. If the past is any guide, the future may provide us with some interesting developments. It would be idle to speculate in any detail just what those developments might consist of, the engine having proven highly resistant to prognostication until now. But we may look for a clue in the milieu in which the engine was brought to its present state, to see if there are any pointers to how things might turn out.

It seems likely that some of the future developments of the engine will involve the big battalions. We're not talking only of capital here, but of government – for although both the free-piston and the kinematic engines were developed by independent companies, the impetus and the funding which brought success came from government, if at a few removes. And we should remember that the failure of solar Stirling so far is because of the intervention of a government – in this case the Chinese – whose massive state investment brought the cost of photovoltaic down to a level at which Stirling could not compete. That coin has two sides, though, and it may be that in future a US government, freed of the trammels of neoliberal economic orthodoxy, might see its way to taking up William Beale's suggestion of using a 170-mile square of the Nevada desert to supply the US with cheap, clean electricity. One can only hope that something of the sort might happen in time to make the building of a 5,000-mile-long sea wall unnecessary. The technology exists and has been proven. All that is wanting is the political will. And a lot of cash.

The crystal ball is a bit dim as regards the submarine kinematic Stirlings, mainly because so little of the knowledge gained from their research has escaped the confines of military secrecy. But it isn't too hard to envisage a place for a highly efficient machine which at a certain scale converts heat into motion – and does so quietly. But on most prospects for the Advanced Stirling, it must be said that the free-piston engine is what appears. The crystal ball is not short of views of how the FPSE might be used to advantage – with one proviso: that it be capable of being turned out reasonably cheaply in large numbers. Given past performance, we may expect to see this happen, provided the incentive exists – and for that the intervention of government may be required.

Both John Malone and William Beale could see a place for small-scale communal power generation using the FPSE. Malone envisaged a generator at the end of every street and Beale speculated on the advantages of an FPSE-powered cogeneration unit fired by carbon-neutral renewable fuels. The backyard heat and power plant is a very attractive idea to a great many people, especially in areas of low population density. Cheap, long-lasting maintenance-free FPSEs in this guise may yet make a significant contribution to human welfare and environmental conservation. Between photovoltaic and windpower generation, there may be a space in which individuals and small organisations may use the FPSE to make a significant contribution to the generation of electricity. Government can be effective in this by providing incentives such as tax breaks and subsidies, which will enable small ventures to succeed where without them they would not.

It may be apparent that so far we have made little mention of nuclear fission reactions as a possible terrestrial source of the heat with which to run an FPSE. The idea is not new: indeed, it is almost as old as the FPSE itself. In 1967, an application for patent for a generator was lodged by E. H. Cooke-Yarborough at the Harwell Laboratories of the United Kingdom Atomic Energy Authority. Not only did Mr Cooke-Yarborough envisage a nuclear reactor driving a Stirling engine, but the engine which he described was a free-piston engine and the electricity was produced by a linear alternator. Whether the ideas owed anything to William Beale is unknown (though William was pretty free with his ideas and seemed happy for other people to adopt them). The ideas are all there in the patent application but the invention was not pursued: testament perhaps to the difference in culture between Harwell and Athens, Ohio. Beale was full of ideas about how the free-piston engine might change the world, but we find surprisingly few mentions of nuclear fission as a possible source for the heat to run the engine until NASA came along with a demand which could be satisfied with nothing less. An omission as large as this cannot have been accidental and must have been a matter of policy. If it was, it would fit in with what we know of Beale's views about society, ecology and industry.

You may have noticed that, throughout the preceding

pages, the matter of scale keeps cropping up. What works for the Stirling engine on one scale does not on another. For most of Beale's working life, the generation of heat by nuclear fission reaction was thought possible only on the largest of scales – which may have been a motive in his avoidance of it. NASA's development of small fission reactors for the KRUSTY project marked a step-change. Not only were tiny reactors designed, but they were built and they worked and they didn't give rise to any insurmountable problems. Quite recently, we have seen a rise, in the USA and elsewhere, of interest in small modular reactors: nuclear generating sets which are capable of being shipped on a flatbed truck and installed where they are wanted. One of these days, my crystal ball tells me, we might see them equipped with free-piston Stirling generators. That would make a difference.

It is now some 205 years since Robert Stirling applied for a patent for his engine, and there can be few scientific breakthroughs which have taken so long to reach maturity, proved so complex to understand, so difficult to make work. It is even more apparent these days that Stirling based his understanding of a heat exchange engine on a profoundly wrong theory of heat and only intuited how some aspects of the engine might operate. Yet these are details in the full extraordinary scale of what he turned out to have created. Visionary after visionary sensed the scope of his creation yet all failed – let down by inadequate technology, accident or sheer bad luck. It took until the end of the twentieth century for another genius to finally realise what would make the engine work. Its maturity arrives in a world still in thrall to vastly inefficient and filthy methods of producing power. Not only could it take us towards the stars which Stirling so avidly observed in his last years, but it offers a chance to heal a planet devastated by climate crisis and over consumption. Stirling's gift to us, his far descendants, was not simply that of an engine. It was the gift of hope.

Select Bibliography

Anon. *The History of Galston Parish Church*, Alex Gardner, 1909

Associated Press, 'US to Buy Pu from Russia', 29 December 1992

Atkins, Peter, *Thermodynamics*, Oxford, 2010

—'The Second Law', *Scientific American*, 1984

Baine, *Directory of the County of York*, 1823

Barry, P. L., and W. T. Beale, 'Free-piston Stirling engine conceptual design and technologies for space power', NASA, 1990

Beale, Dan, 'Lean and data-driven', Memorial service remarks, 2016

Beale, W. T., *Free-Piston Stirling Engine: 20 Years of Development*, Sunpower Inc.

—'Bottom-line Thinking', *New York Times*, 15 November 1989

—'Solar-powered Stirling Engines and their Potential Uses in Developing Countries', for UN Industrial Development Organization, April 1986

Beattie, Frank, *Proud Kilmarnock*, Fort, 2002

Berchowitz, David, *Some Reminiscences of Sunpower*, 1997

Berman, Morris, *Social Change and Scientific Organisation*, Cornell, 1978

Bernstein, H.T., *J. Clerk Maxwell on the History of the Kinetic Theory of Gases*, Isis, Vol. 54

Billings, Lee, 'NASA Struggles Over Deep Space Plutonium Power', *Scientific American*, 10 September 2015

Birse, Ronald M., *Engineering at Edinburgh University*, EUP, 1983

—*Science at Edinburgh University, 1983–93*, EUP, 1994

Black, Joseph, *Lectures on the Elements of Chemistry*, Creech, 1803

Brian, Berger, 'Russia Withholding PU NASA Needs', *Space News*, 11 December 2009

Brown, Callum, *A Social History of Religion in Scotland Since 1730*, Methuen, 1987

Brown, G. I., *Count Rumford*, Sutton, 1999

Brown, S. & M. Fry (eds), *Scotland in the Age of Disruption*, EUP, 1993

Burrleigh, J. H. S., *A Church History of Scotland*, OUP, 1960

Burton, Anthony, *Richard Trevithick*, Aurum, 2000

Carnot, S. *Reflexions sur la Puissance Motrice du Feu*, 1824

Carroll, Sean, 'Warp Drives and Scientific Reasoning', web post, 26 May 2015

Cayley, Sir George, letter, *Nicholson's Journal*, 25 September 1807

Cayley, Sir George, 'On Aerial Navigation', *Nicholson's Journal*, 6 September 1809

Church W. C., *Life of John Ericsson*, Sampson, 1890

Cole, Michael, 'Stirling Test at NASA Glenn', NASA, 30 July 2018

Cooke, Conrad W., 'On Wenham's Heated-Air Engine', Proceedings of the Institution of Mechanical Engineers, 1873

Cummings, A. & T. Devine (eds), *Industry Business and Society in Scotland since 1700*, John Donald Publishers, 1994

Davie, G., *The Democratic Intellect*, EUP, 1961

Diako, Anatoly, 'Disposition of Weapons-grade Plutonium in Russia', Center for Arms Control, 1999

Dickinson, H. W., *James Watt*, David & Charles, 1936

Dickinson, H. W. & A. Titley, *Richard Trevithick*, CUP, 1934

Duffett, Derek, 'Stirling Engines' in *The Piston Engine Revolution*, Newcomen Society, 2012

Edwards, Rob, 'Plutonium for Sale', *New Scientist*, 2001

Everitt, C. W. F., *James Clerk Maxwell*, Scribners, 1975

Fairlie, Gerard & Elizabeth Cayley, *The Life of a Genius*, Hodder, 1965

Filipov, David, 'Russia Suspends Pu deal with US', *Washington Post*, 3 October 2016

Fitzgerald, G. F., *Lord Kelvin*, Royal Collection Trust, 1899

Fry, M., *Patronage & Principle*, Aberdeen University Press, 1987

Gauldie, Enid, *The Scottish Country Miller*, John Donald Publications, 1981

Gibson, Marc A. et al. 'Development of NASA's Small Fission Power System', American Institute of Aeronautics, 2017

Hawkins, N., *New Chatechism of the Steam Engine*, Lindsay, 1904

Higgins, Andrew, 'Plutonium "Leaking" on to Black Market', *The Independent*, 17 September 2011

Higgins, A. & S. Crawshaw, 'Russia's Nuclear Car-boot Sale', *The Independent*, 20 August 1994

Jenkin, Fleeming, *Papers: Literary, Scientific, &c.*, Longmans, 1887

Joule, J. P., 'On the Mechanical Equivalent of Heat', *Philosophical Transactions*, 1850

Kelly, Michael, *Steam in the Air*, Pen & Sword, 2006

Kilmarnock District History Group, 'Origins of Kilmarnock: Part 3', *Kilmarnock Standard*, Stirling obituary, 8 June 1878

Kuhn, T. S., *Structure of Scientific Revolutions*, Chicago, 1962

Leslie, John, *Experimental Enquiry into the Nature and Propagation of Heat*, 1804

Mabon, Basil, *The Man Who Changed Everything*, Wiley, 2004

Low, D. A. (ed), *Critical Essays on Robert Burns*, Routledge, 1975

Mackey, Michael C., *Time's Arrow*, Dover, 1992

Marsden, Ben, *Watt's Perfect Engine*, Icon, 2002

—*Engineering Science in Glasgow University*, PhD Diss, 1992

—'Blowing Hot & Cold', University of Kent at Canterbury, 1998

Marsden. Ben & Smith, Crosbie, *Engineering Empires*, Palgrave Macmillan, 2005

Maxwell, J. C., *Theory of Heat*, Longmans, 1875

May, Charles Paul, *James Clerk Maxwell*, Franklin Watts, 1962

Mayr, Otto, *Philosophers & Machines*, Science History Publications, 1976

Mead, Carver, 'The Spectator Interview', *American Spectator*, October 2001

Miotla, Dennis, 'Pu-238 Production Alternatives', Nuclear Energy Advisory Committee, 21 April 2008

Mosher, Dave, 'Warp Drives', *Business Insider*, 26 May 2016

NASA, 'Deep Space 1 Ion Propulsion System', Glenn Research Center, 2008

NASA, 'Demonstration . . . power', press release, 2 May 2018

NASA, 'Ion Propulsion', press release, 7 August 2017

NASA Glenn, 'Large Free-Piston Stirling Engines', press release, tec.grc.nasa.gov

NASA Glenn, 'Advanced Stirling Converter', press release, 19 July 2016

NASA Glenn, 'High Power Linear Alternator', press release, 4 June 2013

NASA Glenn, 'High Power Stirling Test Rig', press release, 17 November 2013

NASA Glenn, 'Nuclear Systems Project', press release, 19 July 2016

NASA Glenn, 'Thermal Energy Conversion', press release, 30 October 2017

NASA Glenn, Marc Gibson, Kilpower Press Conference, 02 May 2018

NASA Los Alamos, 'Nuclear design', press release

NASA Los Alamos, 'Reactor overview', press release

NASA Multimedia, 'Is warp drive real?', press release, 7 August 2017

New Atlas, 'US restarts PU-238 production', 23 December 2015

Nicholson, 'Nicholson's Journal', The Guardian, online index

Organ, Allan J., Stirling Engine Science, ASME, 2020

—The Regenerator and the Stirling Engine, MEP, 1997

—A Re-appraisal of the Stirling Engine, For the right reasons community print, 2016

—with Theodore Finkelstein, Air Engines, ASME, 2009

Paul, David Mel, John Ericsson, Chandler Lake, 2007

Penrose, Harald, An Ancient Air, Airlife, 1988

Pigot & Co., 'Galston', Ayrshire Directory, 1837

Pritchard, J. Laurence, Sir George Cayley, Max Parrish, 1961

Rankine, W. J. M., A Manual of the Steam Engine, Griffin, 1888

—'On the means of realising the advantages of the air engine', Edinburgh New Phil Journal, pp.120–41, 1855

Reid, Dr J. F., Stirling Stuff, Aberdeen University, 2018

'Robert Stirling, Engineer', letter, Irvine Valley News, 12 March 1886

Robertson, James, Agriculture of Perth, Nichols, 1794

Robinson, Eric & Douglas McKie, Partners in Science, Harvard, 1979

Roblin, Sebastian, 'Sunk', The National Interest, 24 February 2020

Rogoway, Tyler, 'Sweden has a Sub', Jalopnik Reviews, 23 October 2014

Romanelli, Alejandro, 'Alternative thermodynamic cycle for the Stirling engine', American Journal of Physics, 85(12), December 2017

Ross, Andy, Solar Engines, Phoenix, 1977

Rowell, Brent, *Bores & Strokes,* Coolspring, 2017

Rumford, Count, 'An inquiry concerning the Source of Heat which is excited by Friction', *Philosophical Transactions,* 88, 1798

Russell, A. R., *Lord Kelvin,* T. C. Jack, 1912

Saraf, Nadini, *James Watt,* Ocean, 2008

'Robert Stirling Obituary', *The Scotsman,* 8 June 1878

Senft, James R., *Stirling Engines,* Moriya, 1993

Sharrock, W. & R. Read, *Kuhn,* Polity, 1998

Shelton, Mark L., *The Next Great Thing,* W. W. Norton & Co., 1994

Shi, Katy, 'Use & Re-supply of PU-238 in the US', Stanford University, 18 March 2018

Sier, Robert, *Hot Air Engines,* Argus Books, 1997

—*Rev Robert Stirling DD,* L. A. Mair, 1995

—*J. F. J. Malone,* L. A. Mair, 2008

—*Hot Air Caloric & Stirling Engines,* L. A. Mair, 1999

—*History of the Fuel Economiser,* L. A. Mair, 1995

Slaby, Jack G., 'Free Piston Technology for Space Power', NASA Technical Memorandum 101956, 1989

Slaven, Anthony, *Development West Scotland 1750–1960,* Routledge, 1975

Smith, Crosbie, *The Science of Energy,* Athlone, 1998

Smith, C. & M. Norton Wise, *Energy & Empire,* CUP, 1989

Smith, James V., *The Watt Institution,* Abertay, 1977

Smout, T. C. *A History of the Scottish People,* Fontana, 1969

Sootin, Harry, *Michael Faraday,* Blackie, 1954

The Statistical Account of Scotland XV, Perth, 1837

The Statistical Account of Scotland XLI, Perth, 1843

The Statistical Account of Scotland XXXIX, Methven, 1843

The Statistical Account of Scotland, Ayrshire, Blackwood, 1842

Stevenson, R. L., *Memoir of Fleeming Jenkin*, Chatto, 1909

Stuart, Robert, *Descriptive Steam Engine*, Nonsuch, 1824

Taylor, Gordon, *Free-Piston Stirling Engines*, International Power Generation, 1979

'Thomas Morton', Science & Invention: FutureMuseum. co.uk

Thomson, James, 'Note of interview with Robert Stirling', 28 April 1848

Thomson, William, Lord Kelvin, *On the Dynamical Theory of Heat*, Transactions of the Royal Society of Edinburgh, March 1851

Turnock, David, *The New Scotland*, David & Charles, 1979

University of Edinburgh, *Charters, etc. 1583–1858*, Oliver & Boyd, 1937

Walker, G., *Stirling Cycle Machines*, Clarendon, 1973

Walker, G. & J. R. Senft, *Free Piston Stirling Engines*, Springer Verlag, 1985

Weightmen, Gavin, *The Industrial Revolutionaries*, Grove, 2007

Williams, L. Pearce, *Michael Faraday*, Chapman & Hall, 1965

Thanks . . .

My thanks to all of the many people who have helped with this book. When I came to make a list, I realised that of course I was going to forget someone, which would be unfair, so I have settled for a general expression of my indebtedness. If you were one of the folk who helped, you'll know it's for you – and if you were one of the very few who quite unnecessarily impeded the effort, I don't suppose you'll notice anyway.

Index

1816 engine 35, 36, 42, 43
1827 engine 57, 58
1840 engine 58–60
Advanced Stirling Convertor 199
aeroplane 71, 72, 75
air independent propulsion 143, 156
American Motor Corporation 153
astronomy 31

Beale, David 159
Beale, William T. 157–78, 221
Berchowitz, David 163
Black, Joseph 17, 54
Boltzman, Ludwig 54
Braithwaite, John 90
Burns, Robert 32, 33, 212

Caloric engine 92–94, 212

Caloric ship (John Ericsson) 98–100
Caloric theory 45–48, 53, 56, 62, 99, 104, 106, 113
Carnot, Sadi 157
Carrol, Sean 209
Cayley, Sir George 66–82
Chevrolet 54
Chinese photovoltaics 222
cholera 217
Clausius, Rudolf 54, 55
climate 7
Cloag Farm 10
combined heat and power 176, 177
Cooke-Yarborough, E.H. 224
Cornelius DeLamater 96
cryocooler 178–80
Cullen, William 16

Davy, Sir Humphry 75, 106

displacer *see* economiser/
 regenerator/displacer
Drummond family 13, 14
Dumbarton Presbytery 23, 27

economiser/regenerator/
 displacer 36, 43, 44, 49,
 51, 57, 58, 61, 63, 158,
 171
Edinburgh 15
Edinburgh engine 37–39, 49
Edinburgh University 16
electric motors 109, 110
Ericsson pumper 111, 112
Ericsson, John 83–103
exotic technologies 209, 210

Faraday, Michael 94, 95
Ford Motor Corporation 152
free-piston Stirling engine
 165, 169–71
furnace gas engine 80

Galston parish 213, 216
Gardners diesels 125–27
General Motors 150–51
Girdwood & Co. 114
Glasgow engine 40, 41, 49
Glasgow University 21
Gotland submarine 139,
 144–45, 155–56

Gurney, Sir Goldsworthy 78,
 79

heat exchanger 9, 10, 34
heat pipes 202
high-pressure steam 23, 29
Holst, Prof. G. 147
Hume, David 16, 17

Industrial Revolution 21
Innerpeffray Library 13
internal combustion engine
 109, 110
ion drive 192, 206–11

jellyfish 145
Johnson, Dr Samuel 15
Jost fan 121–124
Joule, James Prescot 47, 54,
 106

Kilmarnock 27, 28, 31
Kinetic theory 45
Kirk of Scotland 6, 8, 9, 19, 24
Kockums 155, 174
KRUSTY 195, 201
Kuhn, T.S. 105

Laigh Kirk, Kilmarnock 31
Leslie, John 18, 19, 39, 40
liquid Stirling 128–37

Mackinlay, Rev. James 32
Malone, John 128, 137
Maxwell, James Clerk 55
Meijer, Jan 154
Merrimack 86, 87
Methven 216
Monitor 87
Mount Tambora, volcano 7
Morton, Thomas 30

NASA 152–56, 174, 184–91
NASA Advanced Radio-
    isotope Generator 199
NASA ASRG 199, 200
NASA Evolvable Mars
    Campaign 202
NASA Flattop Critical
    Assembly 201, 202
NASA Glenn Research
    Center 153, 174, 198
NASA GPHS 198
NASA Kilopower Project 203
NASA KRUSTY 202, 203
NASA Nuclear Power
    Assessment Survey 202
NASA Planetary Science
    Decadal Survey 201
NASA Rhessi Probe 179, 180
NASA RPS 198
NASA Space Electric Rocket
    Test 206

Newcomen engines 22
Nicholson's Journal 66, 67,
    74, 75

Ohio University 157, 162
oil crisis 154
Organ, Allan 48
patents 34
perpetual motion 52
Perth Academy 13
Philips Technical Review 148
Philips, Eindhoven 146, 147,
    152
photovoltaic cells 174, 175,
    196
Playfair, John 18
plutonium 238, 197–201
Portland, Duke of 25, 28, 29
Portobello Road 1
Pountain, Dick 1
Princeton 84, 85

railways 29, 30
Rainhill Trials 90
Rampen, Win 49, 62
Rankine, W. McQ 54, 107
regenerator *see* economiser/
    regenerator/displacer
rhombic drive engine 149,
    150
Rider engine 116–18

Rider, A.K. 114–16
Rinia, H. 148, 150
Robertson, William 17
Robinson engines 125
Robison, John 17
rocketry 205, 206
Rowell, Brent 114
Russia 183, 184

Savannah River 119
scale 224, 225
science fiction 205–10
scientific method 162
scientific theory 47, 48
screw propellor 83
Senft, J.R. 220
Shelton, Mark 169
Sier, Robert 132
small modular reactors 225
solar Stirling engines
101–02, 172–73
Sorensen, Prof. Harry 161
St Andrews University 218
steam engines 214
Stephenson, Robert 29
Stirling Energy Systems 174
Stirling family 11, 12
Stirling, James 57–59
Stirling, Robert xii, 7, 33, 45,
48–53, 56–58, 60–63,
212–21

Stockton, Robert F. 84, 85
Strathbogie 6, 219
Stringfellow, John 77
Sunpower Inc. 166–70, 198
supercritical fluids 132, 133

technology 163
theology 26, 29
thermodynamics 53–56
Thompson, Benjamin 46–47,
75
Thomson, James 50–53
Thomson, William 41, 49,
50, 106
thought experiment 56, 158
Trevithick, Richard 29, 75, 90

United Stirling 152, 153,
155, 173
Urieli, Izzi 176
US Navy 83, 87, 160
USS *Ronald Reagan* 138–46

Voyager 1 xi, 194

warp drive 208
water pumps 111
Watt, James 17, 22
Wenham, J.F. 108